Dreams
TO
BEAMS

A Guide to Building the Home You've Always Wanted

by
Jane Moss Snow
with the
National Association of Home Builders

Dreams To Beams:
A Guide To Building The Home You've Always Wanted

ISBN 0-86718-325-X

Library of Congress Catalog Card Number 88-62039
© 1988 by the National Association of Home Builders
of the United States

For further information, please contact:

NAHB Bookstore
15th and M Streets, N.W.
Washington, D.C. 20005
800/368-5242

Library of Congress Cataloging-in-Publication Data

Snow, Jane Moss.
Dreams to beams.

Bibliography: p.
1. Architect-designed houses—Design and
construction.   I. Title.
TH4812.S68   1988   690'.837   88-62039

ISBN 0-86718-325-X

12/88 Scott/Carter 5M
8/91 Scott/AGS 2.5M
8/94 AGS 1M

CONTENTS

ACKNOWLEDGEMENTS

Jane Moss Snow is the author of *Kitchens, Bathroom Design,* and *A Family Harvest.* She writes extensively for magazines and newspapers. Her work has appeared in *Mortgage Banking, Builder, Parade, Washingtonian,* and *National Geographic,* as well as through Gannett News Service and Maturity News Service.

Ms. Snow has broad experience as a home builder, an interior designer, and in the insured home warranty field.

The author wishes to thank the following individuals for their technical review and assistance in the preparation of *Dreams to Beams*:

Debra Bassert, NAHB Land Use Planner
Susan D. Bradford, NAHB Director of Publications and *Dreams to Beams* editor
Robert J. Corletta, AICP, NAHB Director of Multifamily and Income Property Finance
John N. Drake, President of Drake Homes, St. Charles, Missouri and 1988 Chairman of NAHB's Custom Single Family Builders Committee
Glenna Etherton, NAHB Editorial Secretary
William Ethier, NAHB Litigation Counsel
Robert Gluck, NAHB Remodelors™ Council Information Specialist
Curt Hane, NAHB Assistant Publications Director/Editorial
Kristine M. Holland, NAHB Director of Sales and Marketing
J. Michael Luzier, NAHB Assistant Director of Land Use and Development Services
Richard Morris, NAHB Codes and Standards Specialist
Michael F. Shibley, NAHB Director of Land Use and Development Services
William Young, NAHB Director of Consumer Affairs/Public Liaison

Dreams to Beams: A Guide to Building the Home You've Always Wanted was produced in cooperation with NAHB's Custom Single Family Builders Committee and the NAHB Consumer Affairs Subcommittee, under the general direction of Kent Colton, NAHB Executive Vice President, by James E. Johnson, Jr., Staff Vice President of Information Services; Adrienne Ash, Assistant Staff Vice President of Publishing Services; Susan D. Bradford, Director of Publications and *Dreams to Beams* editor; and David Rhodes, Art Director.

The house of every one is to him as his castle and fortress.

Sir Edward Coke
(1552-1634)

INTRODUCTION

*Y*our home should be your castle—the house of your dreams. Whether it's your first home or the home you've waited a lifetime to build, your dreams become reality when you have a home custom built.

Dreams to Beams: A Guide to Building the Home You've Always Wanted takes you step-by-step through this exciting and very personal process. By understanding the planning, design, and building of a house, you can shape the outcome of your dreams. The result: a home you will be proud of, and one that satisfies your needs.

Chapter 1, Planning Your Dream House, helps you to decide where you want to live, what type of house you want, and what features will make this your dream home. The chapter contains extensive planning and design checklists for all areas of the home.

In Chapter 2, Working With Architects, Builders, and Interior Designers, you will learn how to select the team to design and build your home, and how to achieve optimum results in working with them. The chapter discusses each professional's role on the team, plus finding the right person, fees, warranties, and how to read blueprints.

Chapter 3, The Land, tells you how to select a specific lot for your dream house. The chapter covers zoning, covenants, easements and encroachments, as well as the physical aspects of the site you are considering. The chapter includes a site selection checklist.

Chapter 4, Financing, helps you determine how much you can afford to spend, with comprehensive financial profile worksheets that you can fill in yourself. The chapter covers financing for land acquisition, construction, and the mortgage, with detailed information on sources for each.

Chapter 5, Cutting Costs, offers options for saving money if you find you are over budget, from postponing certain finishing touches to do-it-yourself possibilities.

Chapter 6, Contracts and Insurance, details the various types of contracts and insurance that are required for a residential construction project.

Chapter 7, How a House is Built, walks you step-by-step through the construction process, from foundation to finish. The chapter also covers codes and permits, inspections and change orders, and features a typical construction timetable for each phase of the homebuilding process.

What if you plan to remodel your current home to create the home you've always wanted? Chapter 8, Remodeling Your Home, focuses on those aspects of planning, design, and construction that are unique to remodeling. The chapter addresses the remodeling team, financing, contracts and insurance, and life during remodeling.

In Chapter 9, Landscaping, you will find out to work with landscaping professionals and how to plan a landscape design that enhances your dream home. The chapter also covers maintenance and landscape lighting.

Your dream house is finally finished and you are ready to move in. Chapter 10, Settlements and Settling In, guides you through the last minute details: the presettlement walk-through, warranties, the settlement process, customer service—and of course, move-in. The chapter features a comprehensive checklist to help you prepare for moving day.

Dreams to Beams contains an extensive glossary of terms you are likely to encounter as you have your dream home built, and a list of resources for additional information on the planning, design, and construction process.

In the pages that follow, you will find out just what is involved in having a custom home built. We suggest that you read through the entire book to familiarize yourself with the process. You can then refer to specific chapters as your plans progress.

The time has come to turn your ideas into wood and brick and stone, into sun-filled kitchens and star-filled master suites—time to make your dreams come true.

Chapter One

PLANNING YOUR DREAM HOUSE

*I*f you have been dreaming about having a new home built, saying to yourself, "Wouldn't it be wonderful if . . . ," now is the time to give shape to your thoughts.

As you plan your new house, you must be able to identify what you want in some detail. It is not enough to say four bedrooms, two baths, a large family room, and an eat-in kitchen. Where do you want to live? What architectural style do you prefer? How many stories will your new home be? What personal features will make this house your dream home? This chapter helps you to organize your thoughts as you embark on that most exciting of projects: having a custom home built.

WHERE DO YOU WANT TO LIVE

Unless you've inherited land that you want to build on, you will need to find a site for your dream house. Ask yourself where you will be happiest: city, country, or somewhere in-between. If you are a theater buff and enjoy a city's varied offerings, rolling green pastures may not be for you. If you long for birds and trees and wide-open spaces, this is your opportunity. Be honest with yourself and assess each potential site carefully. Be sure it's what you want to live with—a place where you can put down some roots.

Choosing a neighborhood

The first consideration is location, which includes neighbors, access, services and amenities, taxes, and potential growth. Take a hard look at every aspect of the neighborhood you are considering: public transportation,

1

schools, places of worship, fire and police protection, the crime rate, civic pride, public maintenance of roads, and zoning. You want to be sure that the neighborhood is compatible with your lifestyle; that the services and amenities meet your requirements; that the area is improving, not going downhill; that taxes are reasonable and any future changes at least partially foreseeable. If the area is growing rapidly and new public facilities will be necessary to accommodate the growth, taxes are likely to go up. You should feel confident that this is where you want to invest your tax dollars for better schools and roads, and that your neighbors are also willing to invest in the community.

How do you learn something about the neighbors? Knock on a few doors and just chat. Tell them you are considering buying in the area and want to find out more about it. You can ask about neighborhood services, and gain an overview of the problems and advantages of living there. While you're doing that, you can decide whether you like the people and their attitudes.

Drop in on a PTA meeting at the school. Browse through the local newspaper. Go to a worship service and talk with people during the coffee hour if there is one. Try to attend a community meeting or two. Visit the local supermarket and other retailers.

You may be planning to live in this house all your life, but it is nice to know that the area will probably appreciate. If you should change your mind and sell your house, you will want to recoup your investment and then some.

Use the following checklist to help you select a neighborhood.

❏	Schools and day care	❏	Traffic
❏	Shopping	❏	Noise
❏	Places of worship	❏	Automobile maintenance
❏	Public transportation	❏	Sun and breezes
❏	Medical facilities	❏	Views
❏	Police and fire protection	❏	Privacy
❏	Parks and playgrounds	❏	Convenience to work
❏	Trash and garbage collection	❏	Convenience to family and friends
❏	Taxes	❏	Other

CREATING A DESIGN FILE

If you have been saving pictures and articles that express your ideas for your new house, these can be incorporated into a file to review with the person who designs your home. The following resources will help you put together a design file.

❑ *Library research.* Comb through architecture and design books in your public or local university library (universities with architecture schools may be especially helpful). Do the same with design-oriented magazines, looking through at least a year's worth of several publications that interest you.

Photocopy everything that catches your eye. Create a file for every room, and for windows, entrance halls, garages, porches, landscaping, and other elements that are important to you. Be sure to include examples of exterior designs you like. Remember that custom building allows you to modify features according to your taste, needs, and budget.

You may want to cross-reference your files. For instance, you may like a certain light fixture shown in a living room picture, but want to keep that picture in the living room file. Cross-reference the picture under light fixtures.

❑ *Local survey.* Visit model homes in your area, making specific notes of features you like. Photograph them or pick up illustrated sales literature for your files. Also photograph existing homes to record design ideas you like.

❑ *Friends.* Look at the homes of your friends carefully (and tactfully). Request permission to photograph any features that particularly appeal to you. Ask who designed and built their home. You may want to approach the architect and/or builder about working on your own home.

❑ *Personal notes.* Keep a notepad handy at all times to jot down stray thoughts, ideas, addresses of places you like, and fresh avenues to pursue. Don't worry if an idea sounds silly or far-fetched; it may lead to something exciting.

As you expand your design file, you may change some of your ideas. If the changes are major, take a good hard look at the reason. Keep weeding your files. Make choices between option "a" and option "b" and toss the loser. This discipline will force you to make decisions early in the process, rather than down the road when each change costs money.

EXTERIOR STYLE

By now you've identified the styles you like and added them to your design file. Assess similarities and differences and list those features that appeal to you the most. Note that styles and features vary by region according to climate, availability of materials, and local taste. A southwestern adobe-style house may be difficult and inappropriate to build in Maine; basements are as rare in Arizona as they are common in Ohio.

Use the following checklist to help you identify those exterior features that appeal to you. When the time comes to have your home designed, you will know what direction you want to take and can communicate your ideas to the architect and/or builder.

Finish

Brick

- ❏ Solid color
- ❏ Variegated
- ❏ Comments _____

Wood

- ❏ Siding (horizontal or vertical, textured or smooth)
- ❏ Shingles
- ❏ Shakes
- ❏ Painted
- ❏ Stained
- ❏ Comments _____

Vinyl siding

- ❏ Comments _____

Aluminum siding

- ❏ Comments _____

Stone

- ❏ Cut (sized according to specifications for a particular location in a structure)
- ❏ Field (as found in the ground)
- ❏ Comments _____

Stucco

- ❏ Comments _____

Other finishes

- ❏ Comments _____

Mixed finishes

- ❏ Comments _____

Shape of structure

- ❏ Symmetrical
- ❏ Asymmetrical
- ❏ One story
- ❏ One-and-a-half story (Cape Cod or salt box; see Figure 1)
- ❏ Two story
- ❏ Split entry or split foyer (living areas a half-level up or down from entry to house)
- ❏ Split level (1- and 2-level living combined in a single house)
- ❏ Other _____

Roof type

- ❏ Gable (front or end)
- ❏ Hip
- ❏ Mansard
- ❏ Flat
- ❏ Gambrel
- ❏ Salt box
- ❏ Cape Cod (dormer)
- ❏ Shed
- ❏ Other _____

"Personality"

- ❏ Open, airy, welcoming
- ❏ Private, secluded, inward-looking
- ❏ Elegant, impressive
- ❏ Old fashioned, cozy, homey
- ❏ New, exciting, different
- ❏ Other _____

The exterior color scheme of a home can reflect its personality. In selecting colors, look at various homes to see what appeals to you. Your architect and builder will probably have definite ideas, as will an interior designer. Soft greys and natural stained siding blend with a wooded site or waterfront, while a white home stands out. Victorian-style homes look attractive (and authentic to their period) in colors such as yellows and blues with contrasting trim. If you opt for this style, you may want to take a look at the imaginative colors San Franciscans have used on their Victorians.

1. Roof types.

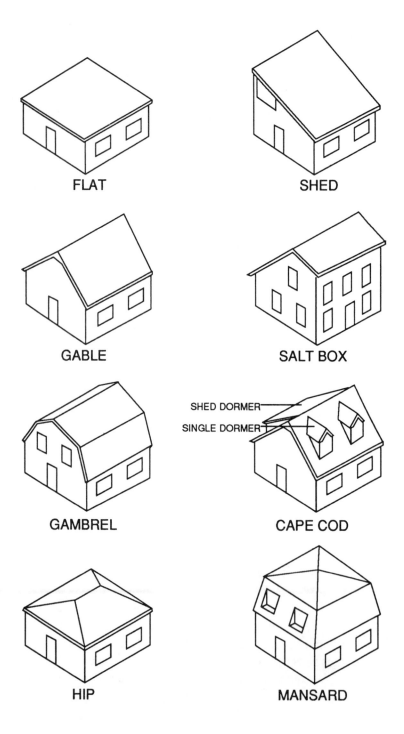

FLAT

SHED

GABLE

SALT BOX

GAMBREL

SHED DORMER

SINGLE DORMER

CAPE COD

HIP

MANSARD

Exterior texture contrasts such as stone with wood or stucco with wood can make a home appear larger and more interesting, if used judiciously. Note that too much texture will give a house a jumbled, poorly designed look. An architect can use contrasting textures or colors to achieve special house-enlarging effects: horizontal bands that carry the eye along the length of the house or vertical patterns that seem to extend the house skyward. Two homes may have the same square footage, but one will seem larger because the design has made the most of texture and color.

YOUR WISH LIST

The difference between your dream castle and just another house rests on those personal touches that make a home uniquely yours. The features you choose should meet your daily comfort and convenience requirements. Keep in mind, however, that if you select features with very specialized appeal, you may limit the number of potential buyers should you decide to sell.

The moment you start thinking of building, begin a "wish list." It may include hundreds of items before you are through, and you may change your mind many times. But having such a list is the best way to ensure that your new home contains the features you've been dreaming about.

Some ideas may occur to you after the walls are up and you see the shape of each room. But changes in plans can be costly if they require structural modifications, additional carpentry, plumbing, or electrical wiring. By including all your wildest dreams in your initial wish list, you can make sensible decisions about which features are essential and which can be added at a later date.

Use the following suggestions to help you get started on your wish list.

Kitchen

❏ Swinging doors between the kitchen and dining room for an informal look; standard doors or sliding pocket doors to close off kitchen food odors and noise.

❏ Open floor plan that brings the kitchen, family room, and dining room together.

❏ Pass-throughs between the kitchen and other rooms for more convenient serving and clean-up.

❏ Eat-in nook or counter, or both, with a large sunny window and overhead lighting designed for the space.

❏ Sliding glass doors or French doors opening onto a patio or deck.

❏ Windows over sink and countertops for sunshine and outdoor views.

❏ View of indoor and outdoor children's play areas.

- [ ] Extra-wide window sills for house plants.
- [ ] Or, if indoor gardening is especially important to you, a greenhouse window in the kitchen for herbs and flowers all year round.
- [ ] Work island or peninsula countertop arrangement.
- [ ] Swing-out wall cabinets for easier access to all shelves.
- [ ] Built-in knife drawer, lazy susan, wine rack.
- [ ] Built-in ironing board.
- [ ] Built-in trash receptacle (how many otherwise attractive kitchens have unsightly waste baskets sitting in a corner?)
- [ ] Walk-in pantry.
- [ ] Broom closet.
- [ ] Direct access from the garage or carport to the kitchen, with a counter close to the door for grocery bags. (Place the refrigerator near the door for ease in putting foods away.)
- [ ] Downdraft stovetop ventilation or hood exhaust system. Each requires special planning, wiring, and venting.
- [ ] Custom cabinets designed to your unique tastes and space requirements (if your budget allows). Mass-produced cabinets are also available in a wide variety of styles, dimensions, and prices.

A word about kitchen cabinets: The standard floor-to-counter height for cabinets is 36 inches, which includes 4 inches for the base and toe space and 1½ inches for the countertop. If you are taller than 5 feet 4 inches, you may want to increase the cabinet height by using a thicker top, adding a shallow drawer, or changing the base height. How do you decide? Try working at a 36-inch-high counter for an extended period of time. If your back starts to hurt, the counter is too low.

Wall cabinets should be hung 15 to 18 inches above the countertop, and the cabinet's top shelf should be no more than 72 inches from the floor for a person of average height. Specify cabinets with adjustable shelves for more flexibility. If the cabinets are hung over a work island or peninsula, they should be set back at least 3 inches from the edge of the countertop and placed 25 inches above it to protect your head from bumps.

Be sure to allow a minimum of 48 inches for clearance between counters. This allows enough room to pass behind someone working at a counter. If two people will work together frequently in the kitchen, try to allow 60 inches between counters.

- [ ] Overhead lighting, positioned to avoid shadowing in work areas. Incandescent lighting generally casts a warmer glow over a room. If you prefer fluorescent lights, choose a warm-tone bulb to flatter food and faces rather than a cold bluish bulb. Fluorescent bulbs may be placed above a dropped ceiling of translucent panels for bright, even overhead lighting. Also specify fluorescent working lights or recessed lighting under all wall cabinets and above the sink, cooktop, and all counters. Overhead and work lights may be supplemented with decorative and accent or task lights that define specific areas of the kitchen.

❑ Floors of ceramic or quarry tile, slate, or marble; countertops and back-splashes of tile, slate, marble, stainless steel, or butcherblock. Consider the additional cost and special maintenance requirements of each. Also consider countertops and backsplashes of durable, easy-to-clean plastic laminate or synthetic marble. These are available in a wide variety of colors, styles, and patterns.

❑ Features that will make life easier for an elderly or handicapped person. These include plenty of floor cabinet storage; 3-foot-wide doorways; countertops, electrical outlets and switches at wheelchair height.

Bath

❑ Whirlpool bath. Plan its placement in relation to the master suite—in a separate room, or as part of the bath or the bedroom. Whirlpools require stronger base support than conventional tubs do, and special electrical wiring.

❑ Separate tub and shower fixtures.

❑ Skylight or glass brick wall for a feeling of openness and light, particularly in an interior bath.

❑ Large wall mirror to make a small bath appear larger.

❑ Bathroom cabinets and vanities adjusted to your height. (Standard cabinets and vanities are 30 inches from the floor.) If the height of the wash basin will change, its plumbing must be adjusted accordingly.

❑ Double wash basins, especially for working couples and families with children.

❑ Vanity to be used for make-up application. Consider specifying a curved surface that is 12 inches deep at center rather than the normal 24 inches. Seeing into the mirror is much easier that way. A vanity or dressing table needs at least 36 inches of clearance to allow a person to pass and a chair to be pushed back. Also consider locating the vanity in a separate area from the tub and commode for privacy and to keep steam from clouding the vanity mirror.

❑ Bright lighting around and above the vanity and wash basin mirrors for well-lit shaving and make-up application (if you use a ceiling fixture as the sole light source, your face will cast a shadow on the mirror).

- [ ] Storage built into the bath area for towels, paper goods, and cosmetics. Adjustable shelves can accommodate items of various sizes; coated wire grid shelving allows air to circulate in a humid environment.
- [ ] Built-in hair dryer rack.
- [ ] Again, features for handicapped or elderly residents, including behind-wall backing for grab bars in the tub/shower and commode areas; 3-foot-wide doorways; lowered plumbing, vanity, electrical outlets and switches.

Master suite

- [ ] Enough room for a quilting loom or set of weights and a television/videocassette recorder (VCR).
- [ ] Or a separate sitting room with plenty of space for loom, weights, a desk and computer, a worktable for hobbies, an easy chair and a television/VCR. This arrangement helps to separate romance and routine.
- [ ] Sunny area or greenhouse window for house plants.
- [ ] Skylight that lets you fall asleep with the moon and stars, awaken with the sun.
- [ ] Large dressing room, or his and hers dressing rooms, with special lighting and built-in cabinets and drawers. Again, adjustable coated wire grid shelving allows air to circulate.
- [ ] Private terrace, deck, or pool opening off your bedroom.
- [ ] Small built-in refrigerator and bar in the master suite.
- [ ] Fireplace. (A self-contained "zero-clearance" fireplace makes this option more affordable than an all-masonry fireplace. This type of unit can be installed directly on wood floors or in walls without creating a fire hazard.)

Children's area

- [ ] Built-ins with shelves and easy-access drawers that can be used for toys and clothes if the children are small, and that can be converted later for books, a computer, a stereo, and sports equipment. Closets should also adjust as children grow.
- [ ] Play/entertainment area that is adjacent to children's bedrooms.
- [ ] If the children will have their own bath, a generously long washing area with two separate basins, and with commode and tub/shower in separate compartments for simultaneous use by girls and boys.

Housekeeping

❏ Laundry room adjacent to the sleeping area, as most laundry accumulates in the bedroom and bath areas. Provide space for ironing and built-in storage; perhaps a cedar closet. Soundproof any laundry area that is adjacent to living space.

❏ Or, a laundry on the first floor or in the basement, with a laundry chute from the bedroom/bath area.

❏ Linen closet that is big enough for extra blankets and oversize comforters as well as bed linens and miscellaneous supplies. Make it accessible from a hall for use by the whole family. If towel storage is not provided in each bath, allow room here. Specify adjustable shelving.

❏ Built-in central vacuum cleaner system.

❏ Mud room for snow boots, beach gear, and gardening supplies. This confines footprints and fingerprints to a single area of the house.

❏ Garage with extra space for storage: built-ins, shelving, or both.

❏ Special built-in cabinets and shelving to display and protect fine crystal, china, or other art objects.

Leisure activities

❏ If gardening is a special passion, a separate gardening room with a stainless steel sink, drainboard, and work surface for potting, plus plenty of shelves and direct access to the yard. Be sure to plan for outside faucets that are adjacent to garden areas.

❏ Swimming pool. These are generally installed by a pool contracting company. Your builder may make the arrangements if he/she has worked with a pool installation company in the past, or may recommend a company with whom you can work directly. Some landscape architects also will oversee swimming pool installation.

Excavation and some preparatory work can be handled when the initial bulldozing is done, provided arrangements are made with the builder. Location of the pool should be considered in siting the house on the lot. Factors to be considered include privacy, sun and wind direction, views, and convenience to indoor and outdoor entertainment areas. An adjacent shower and dressing facilities should be provided.

❏ Tennis court. These are installed by companies specializing in this type of work. Siting should take into account sun and wind direction, distance from the house, and privacy if desired.

❏ Deck with built-in seating, a sunbreak, and a hot tub. (A hot tub requires special wiring and extra support beneath.) Decks add immeasurable leisure living potential to a house, and do wonders for resale value as well. Decks should be incorporated into the original house plan, as their shape and materials affect design. Elaborate or unusual decks require custom work by the builder's crew or a specialist, and material and labor costs will need to be figured into the initial builder bid. Simple decks can be assembled now or later by the weekend do-it-yourselfer with a set of plans or a kit.

❏ Bar in the family room. Should the bar be accessible from the deck or patio?

❏ Built-ins in the family room for a television/VCR, stereo equipment, books, games, a computer.

❏ If a computer is important to you—and especially if you plan to work at home—a custom computer work station.

Interior finish materials

❏ Painted walls. These can be accented with moldings and other trim that range from the contrast of a natural or stained wood to the subtlety of white on cream. If trim is to be stained, let the builder know, as he/she may need to make special preparations beforehand.

❏ Wallpaper, which enriches a room with color, pattern, and texture. Make your selections early in the planning process, as many papers require weeks for delivery and delays could affect the construction schedule. And be sure to alert the builder that wallpaper is to be hung. If wallpaper was not specified originally, it could require a change order that could cost you additional money.

❏ Paneling. Plywood paneling is readily available, but solid wood may require a special order from a lumber yard or home center. Your choice of paneling should be specified in the original plans and included in the builder's bid. If you are using salvage material such as old barn siding, be sure that this is allowed for in the specifications, that you can obtain enough for the walls you want to cover, and that it has been treated for insects. Your builder will probably not be willing or able to track down hard-to-find materials, so you and the interior designer or architect will be responsible for locating it and arranging for delivery in accordance with the construction schedule.

Driveway

A custom home allows you to turn your drive into something other than a straight strip of concrete. Even a relatively small lot can feature a drive that enhances the house. Think of the drive as an introduction. Decide on its location early on so that the bulldozing follows the path you will want in the end. This may save money, trees, and tempers. If you live in a snow-prone area, snow removal and springtime mud will be factors in planning your driveway. You may want:

- ❑ A winding, tree-lined pathway with the house almost hidden until you reach it.
- ❑ The house sited at the head of the driveway so that details reveal themselves as you approach.
- ❑ A curved drive, which can work well on a small lot.

The driveway material is as important as its shape. Use materials that complement the house—brick or cobblestone, for example—and blend driveway and walkway materials where possible. This creates a unified exterior appearance rather than a hodgepodge that distracts the eye. Plan plenty of turnaround space and ample parking for guests and family.

Other possibilities

- ❑ Lighting to illuminate bookshelves, a favorite furniture piece, or a special work of art; floor outlets for lamps.
- ❑ An electronic security system.
- ❑ Passive or active solar features to help reduce energy costs by using the sun's energy to heat indoor air and water. (Passive solar houses are designed to maximize sun energy through the use of simple manual controls such as windows and draperies, shutters, and indoor building materials that retain heat well. Active solar design uses mechanical methods such as solar collector panels and separate heat storage and distribution systems to maximize the sun's energy. Both passive and active solar houses require back-up heating systems.)
- ❑ 36-inch-wide doorways; 42-inch halls; wheelchair-accessible switches, outlets, thermostats; and other adaptable features for disabled or elderly visitors and your own senior years. It costs far more to modify a house after it has been built than to plan adaptability into the original design. Those extra inches can mean added elegance and expanded decorating opportunities—and can enhance the home's resale value.
- ❑ Outdoor wiring for nighttime illumination of landscaping, the facade of your house, sculpture, a fountain, pool, tennis courts.
- ❑ Outdoor phone jacks, if needed.
- ❑ Built-in sprinkler system to keep your lawn and gardens green.
- ❑ Roughed in plumbing and wiring, if you are planning a dream house that will grow as needs require and finances permit.

Chapter Two

WORKING WITH BUILDERS, ARCHITECTS, AND INTERIOR DESIGNERS

With a fairly clear idea of the type of house you want and where it will be built, the time has come to select the professionals who will make your dreams come true.

Most custom home builders are full-service professionals who work closely with their customers throughout the design and construction process. Many custom builders are uniquely qualified to provide their customers with both "design" and "build" services. Or, depending on your budget and design requirements, you may prefer working with an architect to design the home and a builder to perform the construction. Your budget and design needs will also dictate whether you include an interior designer on your team.

Whether you retain an architect or work directly with the builder on the design of your dream home, it is strongly recommended that you involve the builder each step of the way. A builder's input from the start can prevent problems and misunderstandings down the road. Your builder can advise you on design and building approaches, local zoning and code requirements, the latest building products and materials, current costs and ways to save money—all the details consumers face when they decide to have a custom home built.

This chapter defines the role of these design and building professionals and offers guidance on selecting the team that is best for you.

THE DESIGN PROCESS

The primary goal of the person who designs your home—be it a "designer/builder" or an architect—is to satisfy your needs and budget. This means interpreting your ideas, developing the correct relationship between house and site, accurately estimating costs, specifying materials, and providing plans a contractor and subcontractors can follow.

It is wise to select your designer early in the process. He/she can be invaluable in helping you select a site by pointing out both its potential and its constraints and how to make the most of both. As you get to know each other during the land selection process, the designer is better able to understand your housing needs and wants.

Selecting the designer of your dream home will be easier if you understand the basic steps involved in the design process. They include:

❑ *Consultation.* You and the designer discuss design ideas, budgets, timetables, and fee arrangements.

❑ *Preliminary (schematic) design sketches.* The designer researches the site, local codes and restrictions, you and your way of life. This is the time for talking and thinking about land, using the design file you have put together, and explaining each room and your vision of it. The designer will translate your ideas into a set of rough drawings that include site plans (showing the relation of the house to the site), floor plans, cross sections ("sections") and elevations (vertical scale drawings of each side of the house).

❑ *Design development.* Basic structural design is determined. The mechanics of the heating and cooling systems are developed, rooms are laid out, and materials are chosen. A cost estimate is made based on these designs, and adjustments are made to fit your budget. This is the time to make your final design decisions.

❑ *Blueprints or working drawings.* These are the construction plans, drawn to scale, that will be used to build the house. They detail the project floor by floor with perspectives and complete floor plans. Every element is shown, from the placement of light fixtures to the design of the roof overhang.

Blueprints should be signed and dated by both you and the designer to avoid possible future disagreements. Ask how long it will take for the designer to draw up the blueprints. Designers working with computer assisted design (CAD) or existing plans may offer quicker turnaround than those who produce original plans by hand, but you can usually expect to spend 2 to 3 months getting your dreams translated into working drawings.

❏ *Specifications ("specs").* The specifications accompany the blueprints. They are a detailed written list of each type of material and work involved in building the house. The specs identify the quality, type, and brand name or manufacturer of components to be used in each construction phase.

❏ *Model.* At this point an elaborate home may require a model, which usually involves an additional cost.

❏ *Construction bids.* Once blueprints and specs are final and signed, they are ready for construction bids. If you are working with an architect, you may retain him/her to handle this phase, or you can do it yourself. If your builder is designing the house, the builder will often bid on design and construction separately. The blueprints and specifications are the basis both for construction bids and for actual construction. Be sure that the same specs are used for each bid to ensure that all the bids you receive are for identical labor and materials. The bids you receive will tell you whether the job can be done within your budget, or whether adjustments to your plans are needed.

❏ *Construction oversight.* If an architect is designing your home and your contract contains no oversight provisions, you may want to negotiate with him/her to monitor the work as it goes along, consulting with you and the builder throughout the construction process to ensure that plans are being executed correctly. An architect will usually charge an hourly rate for consultation during the construction phase; the fee may range from $30 to $100 an hour and up. Remember, however, that the builder supervises all workers—not the architect.

If your builder is both designing and building the house, he/she will oversee construction.

WORKING WITH AN ARCHITECT

Too often, people are intimidated by the thought of working with an architect. Don't be. As with other professionals, architects provide a service that responds to a particular consumer need.

Your decision to use an architect will be based largely on your design requirements—the type of home you envision for yourself—and your budget. If you know what you want your new home to look like and you have found a custom builder who will bring your dreams to life, you may not need an architect. But if you could use the help of someone with a fresh creative eye and years of design training, an architect can work with you to design the home you've always wanted.

Choosing an architect

One good way to find an architect you like is through word-of-mouth. If you have long admired the home of a friend, colleague, or neighbor, ask about the architect. Chances are your friends will be happy to refer you to their designer (and discuss any problems they had). If you should pass a house whose design you like, stop and ask the owners about the architect. Real estate agents and the local university architecture school are other sources of recommendations.

A list of architects who design residences is available from most local chapters of the American Institute of Architects (AIA). If you cannot locate an AIA chapter near you, contact the national AIA headquarters.*

Membership in the AIA is one way to verify an architect's qualifications. The AIA is a professional society; membership is voluntary. The AIA claims over half of the nation's architects as members. To join the AIA, an architect must be licensed in his/her state. Licensing usually requires an architecture degree from an accredited school of architecture, followed by a three-year apprenticeship and satisfactory completion of a comprehensive registration examination. AIA members use the designation "AIA" after their names. Keep in mind that there are many fully qualified architects who are not AIA members.

When you contact architects, start by asking about their design specialties. If you want a contemporary home, you don't want to hire an architect who specializes in Victorian designs. Interview several architects, preferably in person rather than over the phone. The AIA suggests interviewing 3 to 5 candidates. Some architects will charge for this time, some will include it in their fee if they get the project, and others will give introductory interviews without charge. You will want to ask out about charges before the interview takes place.

Prepare a written list of questions to be discussed at each interview. Your notes will help you to choose among the designers with whom you've met. The architect will also ask you questions about your ideas for the house, your priorities, budget, and so on. How interested are the candidates in you and your project? What priority will they give it? Will there be anyone involved other than the architect you have interviewed?

Does the architect seem to understand your needs? Is he/she willing to discuss your ideas and readily interpret them? Do you feel comfortable with

---

* American Institute of Architects, 1735 New York Avenue, N.W., Washington, D.C. 20006, 202/626-7300.

the architect? The architect will be asking the same questions, as you will need to work well together for the duration of the project. Based on compatibility, design ability, technical competence, and cost, you can make a decision on the architect best suited to design your dream house.

Once you think you have decided on an architect, go out and take a look at the last 5 or 6 homes he/she has designed. Again—talk to the owners and ask for their assessment of the architect's performance. If you are working with a newly licensed architect, you may be one of the first clients. In this case, you will have to judge from drawings and projects produced during apprenticeship.

Architect fees

Architects generally work on a fee basis ranging from 8 to 15 percent of the cost of constructing the house. A good rule of thumb is 12 percent, which means $12,000 on a $100,000 house, $36,000 on a $300,000 house, and so on. Architects also work on an hourly basis, a lump sum basis, a square foot basis, or a combination. A newly-licensed architect may be willing to accept a lower fee in order to gain experience and exposure, while a well-known architect will generally charge more. Fees are negotiable based on the range of services you want and the project's complexity. Your contract with the architect should detail all fees. Architects' contracts are discussed in Chapter 6.

The AIA has published a 16-page booklet, *You and Your Architect*, which supplies additional information on working with architects.*

WORKING WITH A BUILDER

Builders as designers

Some builders have studied architecture and some architects have become builders. Some builders have architects on their staffs or work closely with a particular architecture firm. Many talented building designers are not architects at all, but have developed their design skills onsite as builders. With a keen eye for proportion and detail, nuts-and-bolts knowledge of house construction, and an ability to translate their customers' ideas from words to drawings to finished house, they occupy a unique position in the custom homebuilding market.

If you are considering a builder who has an architect on staff, find out how this arrangement works. Generally the builder constructs the home that

---

* For more information, contact AIA Bookstore, 1735 New York Avenue, N.W., Washington, D.C. 20006, 202/626-7475.

the staff architect has designed. If for some reason you do not use that builder to construct the house, you will generally have to pay a design fee for the plans before you take them to another builder.

Some builders maintain a plan file of homes their firm has already designed and/or built, which they will make available to you. These plans can be modified to your specifications. Many plans are copyrighted, so you will want to check with the owner of the plans to be sure that your use of them does not violate the copyright.

Selecting a set of plans the builder has already used gives you a chance to see what the house will look like. If the builder arranges for you to walk through a house built according to those plans, it becomes easier to see what you like and don't like about the design. Be sure to bring the plans and a notebook with you, and jot down everything you want to question or change. Making changes and decisions in the initial design stage saves time and money later on.

Some builders use computer assisted design (CAD) systems to design homes. CAD offers flexibility and speed, and may allow you and the builder to review proposed plan changes on the computer screen. This does not mean that you can make endless plan changes at no cost, but it does offer an opportunity for experimentation.

If you decide to have your home designed by a builder, establish the terms of the design in a contract. For instance, if the builder finishes the plans and then you decide not to build, have a clear understanding of the costs you will incur. Will you own the plans or will the builder own them? If the builder builds the home, is the cost of designing the plans included in the cost of construction, or is the design a separate fee? Builder contracts are discussed further in Chapter 6.

Find out what kinds of changes can be made in the plans and at what cost. Shifting a door may be a minor change, while adding a laundry room on the second floor may require additional plumbing and an entirely new set of plans—and expenses.

One advantage of working with plans the builder has already designed and built from is that the builder has a specification sheet and some idea of costs. Carefully review the materials the builder has specified in the plans, and be sure you understand exactly what the builder has planned for the house. If your heart is set on a paneled family room and the plans call for painted drywall, this is a specification change and the difference in cost must be calculated. The same would be true for changing a standard bathtub to a whirlpool.

If you work with a "full-service" custom builder, you and the builder can work together to select plumbing and lighting fixtures, flooring, kitchen cabinets and appliances, paint, stain, and wallpaper—the choices that will help to make this home uniquely yours. You may visit the local home supply center together, or make your selections from catalogs and samples in the builder's office. Your builder can direct you to the best values for your budget, and can order specialty items for you.

House plan services

Many people have custom homes built using house plans purchased from a plan service. Thousands of plans are available through plan catalogs and through home and trade magazines. Plans can also be purchased from a number of well-known residential architects who have started selling their own designs in the mass market.

Plans are available for homes ranging from the basic to the large and luxurious. Most plan services use architects to design their homes; most sets come with full specifications and materials lists. Plans vary in price (and quality) depending on the supplier. A full set (including blueprints, specifications, and a materials list for estimating) can be had from a reputable plan service for under $200.

You may shop for plans yourself, or you may prefer to involve your builder in the search. If you find plans that appeal to you, your builder can modify them to suit your particular requirements.

Plan sets offer several advantages:

❏ They encourage comparison shopping because buyers can see complete plans and pictures of what the finished house looks like.

❏ They are far less costly than custom-designed plans would be.

❏ They give the builder a full set of plans to customize according to your requirements.

However, because these plans are "mass produced" rather than being tailored to your specific needs, you will want to be sure that their designs are fresh—not dated. And you will want to check that the plans come with a full set of specifications, including plumbing, electrical, and mechanical. These can be modified according to your needs, the building site, and local codes.

Also, most "mass-produced" plans assume that the house will be built on a level lot. If your lot slopes, your builder will have to adjust the plans accordingly. The resulting house may not take advantage of the site as effectively as one designed specifically for that site.

If you have questions about using a plan service, or if you have found a design you like but are unsure if it can be customized to meet your requirements, your builder will be happy to advise you.

House kits

House kits are complete houses whose construction components have been prepackaged for assembly on the building site. Many homes built from kits are equal in quality to houses built from scratch. Their designs range from the simple to the highly elaborate; many are architect-designed. Most house kits come with a full set of plans for assembly.

Some builders are dealers for a particular kit company, and will work with you to select a design that suits your requirements. The kit company supplies all materials, plans, and specifications and the builder assembles the house.

You may also purchase a house kit yourself and hire a builder to erect and finish it on your lot. As with house plans, information about house kits is available through home and trade magazines, and catalogs. If you choose to buy a house package yourself, you will want to be sure that your builder knows what is included in the kit and what work must be performed. Do not sign any contracts or make any payments until—

☐ you have gotten firm prices from both the house kit company and the builder, and

☐ your builder has met with the kit company's representative to discuss the work to be performed.

Choosing a builder

If you have decided to work with a builder on the design of your new home, chances are you will use that individual to build the house as well. If you are using an architect, or have found a set of plans or a kit for your home, the time has come to select a builder for the all-important job of getting the house built.

Compatibility with your builder is essential, as you will be working together closely for a period of several months. You must respect your builder's professionalism, judgement, and expertise. Without the builder's know-how, those piles of lumber and shingles will never become your dream home.

How do you go about finding a builder? As with architects, word-of-mouth is a good bet. Friends or colleagues will be happy to recommend a builder who has done good work for them. Your local builders association, affiliated with the National Association of Home Builders, can provide you with a list of custom builders in your area.*

---

* For more information about local builders associations, contact the association for your state listed in the Appendix of this book. Or contact the National Association of Home Builders, 15th and M Streets, N.W., Washington, D.C. 20005, 202/822-0200.

If you are working with an architect, ask for a recommendation. A trusted real estate agent is another good referral source. And if you have found homes that you like, find out who built them and talk to the builders.

After you have located several candidates, interview each one and inspect 4 or 5 houses they have built recently. Talk to the owners about their experiences with the builder. Ask about the following:

- ❑ Was the home built according to the agreed upon schedule, or were there any inexplicable delays?
- ❑ Did the builder cooperate in siting the home and preserving landscape features that were important to the owner?
- ❑ Did inspections go smoothly?
- ❑ Was the quality of materials as specified?
- ❑ Were there items that had to be redone because of unsatisfactory work?
- ❑ What about finishing? If there were problems with the finishing touches—paint touch-ups, last-minute shrub plantings, and so on—were these attended to in a timely manner?
- ❑ After move-in, did the builder return to make repairs when called? Was there a callback system and an established timetable for regularly scheduled follow-up service calls?

Also find out whether any complaints about the builder have been filed with the local consumer protection agency or the Better Business Bureau.

If you already have a complete set of house plans, review them and the budget with prospective builders and get their comments. What kind of construction schedule do they offer? Will they handle construction financing? How will the payment schedule work? What kind of warranty is provided?

A word about warranties: Almost all builders offer some sort of written warranty which sets minimum quality standards. Many builders back their own warranties on workmanship and materials, typically for one year. A warranty backed by insurance costs more, but it offers more protection.

For example, the Home Owners Warranty (HOW) Corporation, the oldest and largest new home warranty company in the country, insures the builder's warranty for one year on workmanship and materials and two years on the home's major systems (plumbing, electrical, mechanical, etc.), and gives direct insurance for ten years on major structural defects (such as a roof collapse). If for any reason the builder fails to meet the warranty obligations, the insurance will cover warrantable items over and above a one-time deductible. If a dispute over warranty coverage arises between you and the builder, an impartial third party will mediate. The dispute settlement procedure costs nothing, and it is faster, more efficient, and less emotionally draining than hiring a lawyer and going to court.

There are other insured warranty programs on the market; not all of them offer the same coverage. Whether the builder offers a warranty that is insured or uninsured, ask for details about the service program.

When you are satisfied that you and the builder can work together, ask for a firm bid. You may want bids from more than one builder, but price alone should not be the deciding factor. Quality, reliability, and compatibility are essential.

If the architect will be monitoring construction, be sure that there is good rapport between architect and builder. You don't want to end up in the middle of a personality clash. This is one good reason to deal with architects and builders who have worked together successfully in the past. They operate as a team, and you benefit from their efficiency and cooperation.

Builder fees

Most custom builders will estimate the cost of constructing the house and include their profit in the bottom line at a range of 10 to 20 percent of the total cost. Many buyers prefer this arrangement as it gives them a predetermined total for the job. If changes are made to the plans on which the bid was based, those costs are usually billed separately and are added to the estimate. Additional builder profit is included in bills for changes.

Builders may also work on a cost-plus basis, where all materials and labor are billed at cost and the builder receives a set percentage or fixed fee over cost. Fees may run between 10 and 20 percent over cost. This fee applies to changes as well as the initial plan. For instance, if you decide to upgrade the front door to a more expensive model, the builder's fee will be based on the higher price.

WORKING WITH AN INTERIOR DESIGNER

Imagine moving into your dream house only to discover that your prized antique highboy is two inches too large for its allotted space. An interior designer can help you tailor the interior of your home to your unique requirements. Although many architects and custom builders are skilled at space planning, designing built-in furniture, and recommending finishing touches, the real experts in this field are the interior designers. Whether you use their talents depends largely on your budget.

Minor plan changes can make a world of difference in the finished home. The placement of a skylight to accent a foyer; design of moldings, chair rails, mantels; the addition of 6 inches on a wall to accommodate a 72-inch sofa; the location of electrical outlets and telephone jacks; moving a door a foot to open up a vista—all of these and more are second nature to a designer, and can make your house more beautiful, more practical, more uniquely yours.

The best time to bring in the interior designer is at the beginning when plans are being drawn. Some architects and interior designers work together as teams, particularly on very expensive homes. If you are not planning to have your interiors done professionally, you can retain a designer as a consultant on an hourly basis to work on the initial plans.

As with architects, interior designer fees depend on the individual's reputation, experience level, and geographical location. A recent design school graduate still serving an apprenticeship may charge $25-$50 an hour, while a leading designer could command up to $250 an hour or more. If you plan to use an interior designer throughout design and construction, you may find it more advantageous to develop a cost-plus fee structure or set up a contract based on the wholesale price of the materials plus a consulting fee. Most designers are flexible and will work with you to develop an arrangement that suits your particular needs.

How do you find a qualified interior designer? The American Society of Interior Designers (ASID) is one source. To qualify as professional members of ASID, designers must have six years of experience (education plus an apprenticeship) and must pass a comprehensive exam. Members list "ASID" after their names. Contact ASID for a list of local chapters and members in your area.*

Another way to locate designers—and see examples of their work—is by touring showcase houses. These local fundraising events feature the work of interior designers in your community. Touring them will give you design ideas as well as names.

Designer referral services are now available in some major cities. These match the client with the designer through interviews and slide or video presentations.

The National Kitchen and Bath Association (NKBA) certifies designers who are trained in kitchen and bath design.** These are experts in planning. Non-NKBA-certified interior designers also specialize in kitchen and bath design. It may be worth your while to consult with a specialist in planning your kitchen and bath.

As with the architect and builder, you should have a good rapport with the interior designer. This means being sure that your needs and tastes are clearly understood. Talk to several candidates, look at their work, discuss

* American Society of Interior Designers, 1430 Broadway, New York, New York 10018, 212/944-9220.

** For more information, contact the National Kitchen and Bath Association, 124 Main Street, Hackettstown, NJ 07840, 201/852-0033.

budgets, time constraints, and get a feel for the designer's ability to interpret your ideas.

While interior designers are not essential to your dream home, they can help make a good design great.

THE TEAM

You've selected a builder, an architect (if you are using one), and perhaps an interior designer. Your decisions were based on a combination of compatibility, trust, reputation, and experience. Now you must work together as a team to make your dreams come true. Any plan changes require good communication and prompt decisions among you and your spouse, the architect, builder, and interior designer.

As owner, it is up to you to establish the lines of authority among team members. All players should clearly understand their roles in the design and building process. If you delegate decision-making authority to someone else, be sure that all members of the team know that you have designated this individual to be your representative on the job.

HOW TO READ BLUEPRINTS

Even with all this talent at your fingertips, you still should have a basic understanding of how to read blueprints. Blueprints are the "map" that guide you through your home-to-be. Knowing how to interpret them allows you to spot possible problems and communicate more effectively with the design/construction team.

A set of blueprints usually includes:

❑ Site plan: how your house is positioned on the lot.
❑ Foundation plan and section.
❑ Floor plans of each level of the house.
❑ Elevations: scale drawings of each side of the house.
❑ Building and wall sections: cross section drawings showing structural framing and interior composition.
❑ Special construction details: stairways, fireplace, moldings, and the like.
❑ Mechanical systems: electrical, plumbing, heating and air conditioning.

Blueprints are drawn to scale, usually 1/4 inch or 1/2 inch to the foot. To translate these measurements into something you can relate to, start by looking at the floor plans. Say the master bedroom dimensions are indicated at 16 by 20 feet. If you don't know how big that is, measure it out in your present home. Visualize how much room your bed requires and see what space you have left over. Try this in each room to make sure that the space allocations are satisfactory for your furniture and living needs.

The elevations tell you what the exterior of your house will look like. Look at existing homes and compare them with your plans. If your plan shows a 40-foot facade, find houses of the same dimension to get an idea of the size of your new house.

Blueprints use a "shorthand" of standardized lines and symbols. It is helpful to know what some of these symbols look like, because they indicate materials, door and window types, the placement of electrical outlets* and other details that are critical to the design of your home (Figure 2). These symbols are readily understood by both builder and architect. Any specially designed features are usually shown in a separate detailed drawing.

Don't be afraid to ask questions as you work with the people who will design and build your dream home. After all, they probably don't know much about your line of work. You aren't expected to be an expert in theirs.

---

* The placement of electrical outlets is controlled by code in most communities. Generally, outlets must be positioned so that they can be reached by a 6-foot cord anywhere along a wall that is unbroken by windows or doors. More than likely, you and your builder agreed on outlet placement before the blueprints were drawn up. If not, check the blueprints and let your builder know if you want something other than the standard placement.

2. Common blueprint symbols.

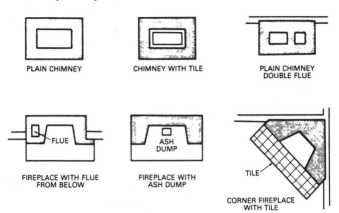

PLAIN CHIMNEY

CHIMNEY WITH TILE

PLAIN CHIMNEY
DOUBLE FLUE

FIREPLACE WITH FLUE
FROM BELOW

FIREPLACE WITH
ASH DUMP

CORNER FIREPLACE
WITH TILE

CHIMNEYS AND FIREPLACES

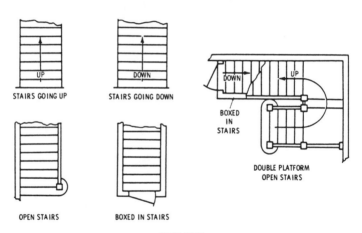

STAIRS GOING UP

STAIRS GOING DOWN

BOXED
IN
STAIRS

OPEN STAIRS

BOXED IN STAIRS

DOUBLE PLATFORM
OPEN STAIRS

STAIRS

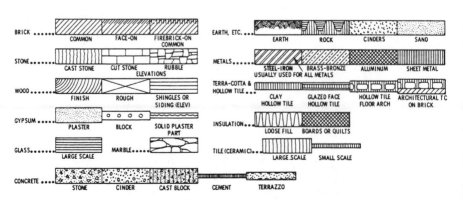

BUILDING MATERIALS

28

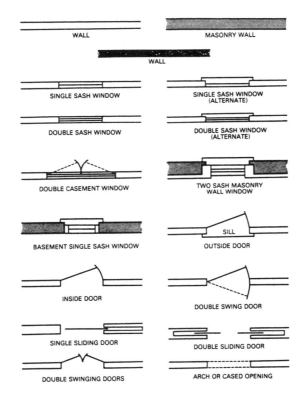

WALLS, WINDOWS, AND DOORS

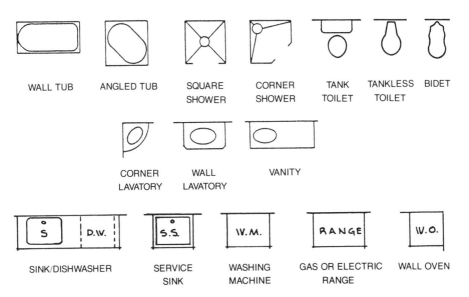

PLUMBING AND MISCELLANEOUS FIXTURES

Chapter Three

THE LAND

*M*any builders offer buyers a "package deal" that includes both a lot and a custom-built house. Some people prefer to buy a piece of land themselves and then retain an architect and/or builder to design and build on the lot. Once you have found a potential site for your dream house, you will need to find out if it is affordable—and buildable. If you have already selected an architect or builder to design your house, have the land checked for buildability and compatibility with the type of house you want.

While it is easy to see the advantages of a particular piece of land, it takes investigation to uncover possible constraints. This chapter covers site planning and the restrictions, requirements, and physical site conditions you may encounter in your search for the perfect lot.

SITE PLANNING

Think about accessibility as you look at a site for your new home. Decide on the best place for a road or driveway so that the builder can get equipment in without great difficulty. If you think you are willing to put up with an unimproved access road for now, how will you feel when you have a foot of snow or springtime mud to contend with? Will you have to install a long electric power line? If you have to put in a long driveway and a long power line to your house, you may be faced with some sizable expenses.

Plot your dream house on the site. Pace it off and put up some temporary stakes. Figure where you want to sit outside, what views the living room should have, where the children can play. Do you want sun streaming into the kitchen in the morning? Are you planning a greenhouse, a garden, a patio, a pool? Do you want to have breakfast on the deck in your pajamas or sunbathe in privacy? Which rooms will you use during the day, and where do you plan to spend most of your time in the winter months? Siting your home for privacy and maximum enjoyment of nature takes a bit of thought. You may have to sacrifice room size or yard space to retain the best view or ensure the greatest privacy.

An energy efficient home will take advantage of the sun for warmth and the prevailing winds for cooling. The local weather bureau has information on the prevailing winds, which your builder or architect will want to check. Facing the street as everyone else does may not be the best angle for your house.

And what about trees? Are they deciduous trees that will shade your house in summer and let in the sun's warmth in winter? Or are they evergreens that will shade your house year-round and provide a block from winter winds? How many trees must be sacrificed to fit the house and drive on the lot? Will you lose all of the big trees and be left with scrub? What does that do to the site? How much will it cost to clear the land for building?

Do you get your magnificent view from a river bank that is eroding away? How much has it eroded in the past five years? Are you willing to gamble that it won't crumble under you? Check erosion data with the local U.S. Soil Conservation Service office. You will also want to know about landslides and earthquakes in high-risk areas such as California.

Try to visit the site in various weather conditions to detect possible air pollution. Listen for traffic sounds and vibrations at rush hour. Are there railroad tracks in the vicinity? How frequently do trains run? Visit during a heavy rain and see what happens to the water. And don't forget about air traffic. How close is the site to an airport traffic lane? Check on waste disposal in the area, industrial pollution of any kind (particularly if the site is on the water), and proximity to chemical hazards.

RESTRICTIONS AND REQUIREMENTS

Zoning

Many cities and suburban municipalities have developed master plans (also called comprehensive plans) that identify future growth areas and general goals for the community's development. These plans are the basis for the more detailed local government zoning ordinances.

Zoning ordinances govern the location and density of various types of land uses, which are generally identified as residential, agricultural, commercial, industrial, and office. Each of these use categories may be divided into subcategories (called zoning districts) that regulate permitted activities and densities or allowable lot sizes in that district. For example, one residential district may be zoned for 5-acre lots, another for ¼-acre lots. Zoning also controls setbacks, which are the minimum allowable distances between a structure and its lot lines.

32

Both master plans and zoning maps are available in many public libraries or from the local planning department. These documents offer clues about development patterns in the community. Your builder, architect, or real estate agent should also be able to advise you on how local zoning requirements will affect the property you are considering.

If you are interested in a particular site, you must determine that zoning allows a private home and other improvements on that lot. Also find out what requirements apply to the location of the house on the site. If possible, you want to avoid having to seek a variance (permission for a use or lot setback that does not conform to a strict interpretation of local zoning regulations). Variance requests are often a lengthy process involving applications and hearings—and the possibility that the request will be denied.

If your future expansion plans call for a room addition, a swimming pool, or a garage, you also want to be sure that there will be enough room on the lot to add them. The zoning may allow a 2,500-square-foot home, which is what you are planning, but the setback requirements might preclude an addition and a tennis court later on.

Are you planning to move your dental practice to your house at some future date? If so, be aware that zoning controls such accessory uses, as well as house height, parking, outdoor signs, and other features. You will also want to check with the local planning office about any local, state, or federal environmental controls that pertain to the site you are considering. These could include septic system and well requirements, tree preservation requirements, wetland protection, flood plain regulations, and the like.

Future plans for the area are also a vital consideration. That beautiful wooded strip across the way may be zoned for commercial use or a highway access road. Ask questions. If you are looking at an undeveloped site in a highly developed area, find out why it's still available. Has it been tied up in an estate or does it lie next to a proposed road?

Examine the applicable local zoning ordinances for the zoning of the parcel you are considering as well as neighboring properties. If you have any questions, check with the local zoning officer or commission, or the local planning board.

Covenants

Covenants (also called deed restrictions) are agreements between a seller and buyer restricting the use of a piece of land. Covenants may also be established by neighborhood associations, developers, or local governments to protect or preserve a desirable element of the community. Find out whether the land you are considering has covenants on it that might prevent you from building a contemporary-style house or adding a fence, or which otherwise affect your use of the lot.

Easements, encroachments, and squatters

Find out whether any neighbor or local government has easements or encroachments on the property you are considering. An easement is a right given by the owner of a piece of land to an individual, government, or other entity for a specific limited use of that land. Easements may be recorded or unrecorded. A recorded easement might involve a utility that has been given the right to run power or sewer lines through the land, or mineral rights, soil removal, and the like. An unrecorded easement might apply to a driveway used by a neighbor for access to adjacent property.

An encroachment is an improvement to property that illegally extends beyond that property's lot line. A neighbor's garage or driveway may encroach on your lot by three feet (or your proposed drive may encroach on someone else's land.) Encroachments cloud a title and must be cleared up prior to purchase of a lot. The encroacher can purchase the land in question, or remove the encroachment, or seek an easement.

How do you find out about easements and encroachments? Start by looking at a recent property survey. The present owner may have one you can check. Also inspect the current deed before you buy. Ask to see the title insurance policy, which covers easements, encroachments, and other restrictions. Your purchase agreement should be contingent on a clear title commitment based on a property survey that covers all easements and encroachments. This survey should be more complete than the survey required by the mortgage lender as part of the loan application.

Another rare but possible problem is the presence of squatters on the property. If people have built a house or shack on the property, they will have to vacate or be evicted. Check with local authorities before taking action, as squatters have certain rights; evictions must be handled according to local law.

Title

Does the current owner hold clear title to the land? The title insurance policy and the trust deed will tell you. When you request title insurance, the insurance company will find out if there are liens or lawsuits on the property which must be resolved prior to your taking title.

Is the property mortgaged? If so, how? If you will be handling construction financing yourself, is there any reason why the property cannot be used as collateral for the construction loan? Until you have your own title search and insurance, you cannot be positive that there is no problem. Any land purchase contract should contain a clause making the sale contingent on a clear title.

If the land has recently changed hands, it may seem ridiculous to have another title search done, but it is the only way to determine what current problems may exist. You may save a few dollars if you shop for a title company that will give a re-issue rate. This is a reduced rate on title searches for properties that have changed hands within the past 5 or 10 years. It is worth asking about.

Interstate Land Sales Act

If you are buying land from a developer who has contacted you from another state or by mail, the property may come under the Interstate Land Sales Act of 1968. This law was enacted to protect the public from fraudulent land sales operations. It applies to subdivisions of 100 lots or more, and requires the seller to provide prospective buyers with a property report containing detailed information about the land being offered for sale. The report should assist you in making an educated decision about the purchase. The law is administered by the U.S. Department of Housing and Urban Development (HUD). HUD does not have any responsibility for the property reports.*

Restrictions and requirements checklist

☐ Zoning district
☐ Future plans for area
☐ Existing zoning variances
☐ Covenants
☐ Easements
☐ Encroachments
☐ Squatters
☐ Clear title
☐ Interstate land purchase

PHYSICAL SITE CONDITIONS

Soils and drainage

Compensating for soil problems can be expensive, so it is best to know about the soil conditions on a lot before you buy. Is the soil sandy, rocky, clay? Is the soil permeable enough to permit water seepage and absorption? Heavy clays, for instance, will not absorb moisture and will cause puddling and flooding. In addition, they can expand and compact enough to stress a foundation, and may require specially engineered supports. Other types of

---

* If you are considering buying land interstate, it would be advisable to read *Buying Lots from Developers [HUD 357-1(6)]*. This U.S. Department of Housing and Urban Development (HUD) booklet is available from the Superintendent of Documents, U.S. Government Printing Office, Washington, D.C. 20402.

clay, such as bentonite, change form and volume dramatically in response to wetting and drying and can be unpredictable soils on which to build. A very weak or sandy soil may also require specially engineered foundations. Find out whether the site is located on fill. A house built on fill will settle and crack at an unacceptable rate unless the foundation is engineered to prevent it.

Will disposal of large boulders be a problem, or can you use them in your landscaping plan? Will a bed of solid rock below the surface require expensive blasting for the basement and utility trenches?

Information about local soil and drainage is usually available from the U.S. Soil Conservation Service office in your county. A soil test or geotechnical survey is not usually required in order to build, but it provides the most reliable information. If you are planning to install a septic system, the county or local health department will require that the soil be tested to determine the rate of water absorption on the site. This is called a percolation or "perc" test.

The topography of the site is also important. A completely level site with nowhere for water to drain can be as much of a problem as a hilly one that threatens to flood a neighbor unless run-off is diverted. Run-off water can also flood your own basement if the house is in the path of run-off or is sited in a depression without an adequate drainage system. Find out who is responsible for storm drains. You? The municipality? The county?

A topographical survey determines elevations that dictate drainage patterns, the pitch of the driveway, slope directions, and possible basement configurations. Many counties have topographical maps, but these may not be up to date. In any case, you may need a current "topo" to get a clear title. It is also one of the first items your architect will want. Topographical surveys are performed by the same firms that do soil tests and other surveys.

If the land is located in a designated flood hazard area, your lender will require you to provide flood insurance. You can check the level of flood hazard through Federal Insurance Administration (FIA) maps. Two types of FIA maps are available. Flood Hazard Boundary Maps indicate general flood boundaries for most flood prone areas of the country, but do not provide specific flood water height or risk information. Flood Insurance Rate Maps are more detailed and include insurance rate tables based on the level of risk; however, these maps are not available for some areas. Check with the local building department, the city clerk's office, or the regional Federal Insurance Administration office* for more information on the level of flood hazard in your area.

---

* For more information, contact the Federal Insurance Administration, 500 C Street, S.W., Washington, D.C. 20472, 202/646-2780.

Utilities

Where will your water come from? A well or the city reservoir? And where will the sewage go? Into a septic tank in your yard or a citywide sewer system? The further from a city, the more likely you are to be beyond municipal utilities.

If you need water, can a well be dug, and how deep must it be? What are typical water flow rates? Ask the neighbors about their experience or consult the health department. Get estimates from two well-drilling companies. Remember that both wells and septic tanks will have ongoing maintenance costs.

Local regulations specify minimum distances between wells and septic systems (both yours and your neighbors'), and require specific amounts of land for septic drainfields. Land with a high water table will not support a conventional septic system, nor will land that has low absorption. Your builder or the health department may be able to recommend an alternative septic system.

If you have access to public services, it is your responsibility (and expense) to have trenches dug up to the house to accommodate electric, gas, telephone, cable television, sewer, and water lines. Some jurisdictions allow "common trenching" of several utility lines in a single trench. Get estimates from the local utilities on the cost of running lines to your lot. If the site is perched on a cliff, the trenching cost may be substantial. There are also hook-up charges for connecting to the various utilities. Find out how far ahead utility hook-ups should be scheduled to avoid construction delays.

Note that rural residents may have to pay for garbage collection and disposal, snow removal, and access road maintenance. Be aware that fire protection may be limited in less developed areas.

If you are choosing between two sites, calculate the raw development costs described above. One site may be more expensive initially, but if it requires fewer utility expenditures, it could be cheaper in the long run.

RADON

Radon is a colorless, odorless, tasteless, radioactive gas that occurs naturally in soil, underground water, and outdoor air. It is found in varying levels throughout the United States. High indoor radon concentrations have been associated with increased health risks. Although soil can be tested for the presence of radon (usually at high cost), no correlation can be made between radon levels in raw land and indoor radon levels in a finished house. For instance, a given lot may have low soil readings for radon but a high indoor reading in the house once it is completed.

Your builder can employ construction measures that prevent radon entry into the house or allow its escape. Many of these measures are common building practice and will not add significantly to the cost of the house.*

Physical site conditions checklist

❑ Soil type and any resulting limitations
❑ Rock
❑ Drainage
❑ Storm drains
❑ Erosion
❑ Flooding
❑ Public water supply and sewer or private well and septic system
❑ Hook-ups to public utilities (electric, gas, telephone, cable television, water)
❑ Trenching costs and restrictions
❑ Radon
❑ Trees and other natural features
❑ Access road
❑ Other

---

* Contact your state health or environmental office or regional U.S. Environmental Protection Agency (EPA) office for further information on radon. EPA has produced the following publications about radon: *Citizen's Guide to Radon, Radon Reduction: A Homeowner's Guide (2nd Edition), Radon Reduction in New Construction: An Interim Guide,* and *Radon Reduction Techniques for Detached Houses: Technical Guidance.*

Chapter Four

FINANCING

*O*nce you have determined what you want your dream house to look like, where it will be located, and who will design and build it, the next step is to obtain financing to pay for it all. Many people are intimidated by the financing process. Don't be. While there are many options to consider, remember that the consumer-lender relationship is based on mutual need. You want the loan, and the lender wants your business. So feel free to ask questions and to shop around.

This chapter helps you to figure out what you can afford to spend and outlines the financing options that are available to you.

HOW MUCH CAN YOU AFFORD

As you shop for financing, you will need to know how much money you can afford to spend on your dream house. The worksheets below will help you to organize your financial picture. Using Worksheet A, you can calculate the amount of money you have available for a downpayment. Worksheet B lists monthly mortgage payments over a 30-year period at different interest rates and loan amounts. Worksheet C helps you to determine what you can afford to spend for housing each month.

Fill in the worksheets carefully. If you are planning to sell your present house to finance your dream home, enter in the estimated amount it will bring after settling the mortgage, paying the real estate agent's commission, lawyer fees, and other expenses. Ask three agents to assess the value of your present house. They will check the selling prices of comparable homes in comparable locations. Don't figure on the top dollar. It is better to err on the low side. List your monthly gross and net income. Be sure to allow for taxes, school fees, dues, insurance and other one-time yearly expenses. Be realistic. You will need these figures when you apply for financing.

Worksheet A: Downpayment estimate

Available funds

Equity in present home .... $ _____

Savings, savings certificates .... $ _____

Investments/mutual funds (current value) .... $ _____

Insurance (cash surrender value) .... $ _____

Other available funds (such as help from family) .... $ _____

Gross total available funds .... $ _____

Minus amount you want to keep in savings .... $ _____

Adjusted total available funds (A) .... $ _____

Expected expenses

Settlement costs (about 5% of home price) .... $ _____

Furniture, furnishings (if needed) .... $ _____

Alterations, repairs, landscaping (if needed) .... $ _____

Moving costs .... $ _____

Other expected expenses .... $ _____

Total expected expenses (B) .... $ _____

Amount available for downpayment

Adjusted total available funds (A) .... $ _____

Minus total expected expenses (B) .... $ _____

Amount available for downpayment .... $ _____

Worksheet B: Monthly mortgage payment

After you determine how much of a downpayment you can make, try to estimate how large a mortgage payment you can afford each month. Your monthly mortgage payment will be determined by the interest rate and amount of your loan. And the amount of your loan plus the size of your downpayment will determine the price of the house you can buy.

Monthly Principal and Interest Payments for a 30-Year Fixed Rate Mortgage at Different Interest Rates and Loan Amounts

Interest Rate	Loan Amount							
	$50,000	$60,000	$70,000	$80,000	$90,000	$100,000	$150,000	$200,000
9%	$402	$483	$563	$644	$724	$ 805	$1,207	$1,609
10%	$439	$527	$614	$702	$790	$ 878	$1,316	$1,755
11%	$476	$571	$667	$762	$857	$ 952	$1,428	$1,905
12%	$514	$617	$720	$823	$926	$1,029	$1,543	$2,057
13%	$553	$664	$774	$885	$996	$1,106	$1,659	$2,212

To estimate how much you can afford to pay for a new home, determine the average monthly income of all the members of your household and then deduct all non-housing expenses. If you are unsure how to estimate non-housing expenses, save and identify all sales receipts for a month or two. From the receipts, your checkbook, and your credit card statements, you should be able to make a reasonable estimate.

You must also compute recurring housing expenses, such as insurance, property tax, utilities, maintenance and repairs. Your lender will consider these expenses when evaluating your loan application.

40

Worksheet C: How much can you spend for housing each month?

1. **Average household monthly income**
   Take-home pay (gross pay less taxes) .... $ _____
   Interest, dividends, rents (do not include resources to be
   used for the downpayment) .... $ _____
   Other income .... $ _____
   Net average household monthly income (1) .... $ _____

2. **Average monthly non-housing expenses**
   Food, household supplies .... $ _____
   Clothing .... $ _____
   Medical costs and insurance .... $ _____
   Life and casualty insurance .... $ _____
   Automobile and insurance .... $ _____
   Commuting .... $ _____
   Installment payments/interest charges .... $ _____
   Recreation/hobbies .... $ _____
   Telephone .... $ _____
   Contributions, dues, fees, etc. .... $ _____
   Personal (dry cleaning, hair styling, etc.) .... $ _____
   Savings/investment program .... $ _____
   Entertainment .... $ _____
   Miscellaneous expenses .... $ _____
   Total average monthly non-housing expenses (2) .... $ _____

3. **Monthly income available for housing**
   Net average household monthly income (total 1) .... $ _____
   Minus total average monthly non-housing expenses (total 2) .... $ _____
   Average monthly income available for housing (3) .... $ _____

4. **Average monthly housing expenses of the home you wish to purchase**
   Principal and interest on mortgage .... $ _____
   Insurance (fire, theft, and flood) .... $ _____
   Property taxes .... $ _____
   Utilities (water, heating/cooling, electric/gas/oil) .... $ _____
   Maintenance and repairs (allow 1% of the price
   of the home per year) .... $ _____
   Other monthly housing expenses .... $ _____
   Average monthly housing expenses (4) .... $ _____

5. **What can you afford?**
   Average monthly income available for housing (total 3) .... $ _____
   Minus average monthly housing expenses (total 4) .... $ _____
   Amount available after paying housing expenses .... $ _____

If total 3 is greater than total 4, you should be able to afford the house. If total 3 is less than total 4, you may have difficulty meeting your housing expenses each month.

FINANCING A CUSTOM HOME

Having a custom home built involves three basic financial steps: land acquisition, construction financing, and permanent financing (the mortgage). You, the buyer, are often involved only in obtaining permanent financing. Your builder may purchase the land and handle construction financing. You obtain a mortgage to purchase the lot and finished house from the builder. In many instances, the builder can arrange the permanent financing for you through his/her lender.

Alternatively, you may find a lot you like and purchase it long before you are ready to build. In that instance, you obtain a loan to purchase the land. When you are ready to build, your builder may handle construction financing and you get permanent financing to pay the builder.

Some custom builders prefer to have their customers obtain the construction loan. And if you act as your own general contractor on construction of the house, you are likely to be responsible for land acquisition, construction financing, and permanent financing.

Land acquisition

Buying land with cash is the simplest way to obtain it, but not always the best way. If you are considering paying cash for a piece of land, consult a tax accountant before you make the purchase. It may be more advantageous from a tax standpoint to use another method for land acquisition and save your cash to cover start-up costs such as architect fees and a retainer for your builder.

One financing method for land acquisition is a signature loan. This type of loan is based on your ability to pay, and is guaranteed by your signature. Your bank decides how much it can lend you based on your income, your credit history, and your record with that bank. If you fail to pay back the loan, the bank may use your personal assets to recoup the money owed.

Signature loans usually carry a fairly high interest rate, but they generally require no collateral and no closing costs. If your lender does require collateral, you will not be able to use the land because a mortgage lender will not give you permanent financing unless you have clear title to the land. Stocks, bonds, real estate, or other tangible assets will serve equally well as collateral. Having a favorable history with the bank will help.

Signature loans are available from commercial banks, savings and loans, credit unions, and consumer finance companies.

Another avenue is to arrange land acquisition terms with the sellers. This arrangement is called a purchase money mortgage. The sellers may agree to accept a downpayment, with the balance due in monthly installments or upon completion of the house and start-up of permanent financing. The land is usually used as partial collateral for the permanent financing and the lender requires the sellers to subordinate their mortgage on the land to the lender. This means that, in case of default, the lender gets first claim on the property and the seller of the land is paid after the lender is satisfied.

Another possibility, and one that many lenders shy away from, is getting a land acquisition loan from the same lender who makes the permanent commitment or mortgage loan. Lenders often prefer to minimize their risk by refusing to finance more than one phase of a project. Although it is more convenient to get both loans from the same lender, it is perfectly acceptable for you to obtain land acquisition financing from one lender and permanent financing from another.

Construction

A construction loan is just that: financing to build the house. It covers material and labor for the actual structure. Note that the construction loan will not cover up-front costs such as the builder's retainer (which may be one percent of the cost of the house or more), architect's fees, property surveys and the like.

Lenders like making loans on something they can see. Lending money for the construction of a custom home means they must take a risk. Many builders have a good relationship with their lenders, however, and obtaining construction financing is a routine part of doing business.

Some builders do not handle construction financing. In that instance, your builder may be able to smooth your path by introducing you to his/her lender. If you are responsible for obtaining a construction loan, the best procedure is to get a commitment for permanent financing first. If you cannot get both permanent financing and the construction loan from the same lender, the permanent financing commitment shows other lenders that you have qualified with a lending institution and are a good risk for a construction loan.

If you are applying to one lender for both construction and permanent financing, that institution will want to qualify you for the permanent financing before committing itself to a construction loan. When the house is finished, the lender simply converts the construction loan into a mortgage. This should save you money in both fees and interest. The land is the collateral for the construction loan; the land and finished house together are collateral for the mortgage.

Alternatively, you may be able to obtain permanent financing at the outset to finance construction. With this arrangement, your lender will require you to make payments on the loan during construction.

Construction loans are made for specific periods of time, based on an estimate of how long construction will take. The interest rate will generally be higher than on other types of loans because of the risk involved.

Most construction loans are paid out over their duration in what are called "draws." Draws are issued according to a predetermined schedule of bank inspections of the construction work in progress. Many banks charge for each inspection, and paid invoices for inspections may be required by the bank for each draw.

Draw schedules vary with the bank and the project. They are usually in 4 to 6 increments, although elaborate houses may require additional draws. A typical 6-draw schedule works like this:

Draw	Stage	Percent
1	foundation	15
2	under roof	15
3	roughed-in plumbing, electric, mechanical	20
4	drywall	15
5	trim	20
6	final	15

Permanent financing (the mortgage)

This is the standard home mortgage, in which the house and land are used as security for a loan. Your options for obtaining permanent financing are described briefly below.

Commercial banks

Commercial banks lend money for a variety of things including land acquisition and construction, but they do not specialize in mortgages on homes. Their primary interest is in shorter term loans.

Thrifts (savings and loans; mutual savings banks)

Thrift institutions have traditionally made mortgage loans. Since 1982, they have been allowed to make other types of loans as well. These institutions are given special incentives and tax advantages if their loan portfolios contain a large percentage of individual mortgages. They may sell their mortgages to other investors, such as the secondary market, and retain the servicing (collection of monthly payments) for which they get a fee.

Mortgage bankers

Unlike other types of bankers, a mortgage banker (or mortgage company) does not accept deposits from individuals or businesses. In making mortgage loans, he/she gathers and processes information on the borrower and the property being financed, funds the loan (if the necessary criteria are met), and then sells the loan to an investor. The mortgage banker retains the servicing.

Mortgage brokers

A mortgage broker does not originate or service loans, but brings together a borrower and a lender, for a fee. You pay for the service only if the broker finds you a satisfactory loan.

Federal Housing Administration (FHA)

The FHA is a major insurer of home mortgages—not a lender. Lenders find this security so attractive that they will accept a lower downpayment from the consumer. FHA currently insures loans to a maximum of $67,500-$101,250, depending on housing costs in the area where the loan is being made.

Veterans Administration (VA)

The VA guarantees loans made to veterans and their spouses. These loans require no downpayment. Although the VA sets no ceiling, VA loans are usually limited by the lender to $144,000. A one percent "origination fee" is usually required up front.

Secondary market

A key player in the mortgage world is the so-called secondary market. To keep the market fluid by providing enough funds to meet demand, primary lenders sell the mortgages they make to other savings institutions, to mortgage bankers, to insurance companies, or to credit agencies such as the Federal National Mortgage Association (FNMA or "Fannie Mae"), the Federal Home Loan Mortgage Corporation (FHLMC or "Freddie Mac"), and the Government National Mortgage Association (GNMA or "Ginnie Mae").

You may ask what the secondary market has to do with you. Plenty. Most individual mortgages are packaged and resold in the secondary market in units of one million dollars or more, so they must meet the standards set by the purchasers. If your loan does not meet specific qualifications, the lender may have to hold the mortgage in its own portfolio. Consequently, lenders prefer to make conventional "conforming" loans rather than "nonconforming" loans that may be difficult or impossible to sell in the secondary market. (A conforming loan is one that meets the secondary market's specific requirements. For instance, a loan will be conforming or acceptable to Fannie Mae if it does not exceed Fannie Mae's maximum mortgage amount and certain other specified criteria.)

TYPES OF MORTGAGE LOANS

All mortgage loans include principal and interest; taxes and insurance related to servicing the loan may be included in the monthly payment as well. When you see any of the letters PITI in the monthly payment quote, they denote the total amount of principal (P), interest (I), taxes (T), and insurance (I), in that order. The two most common types of mortgage loans—fixed rate and adjustable rate—are discussed below. Check with your tax attorney about the tax implications of each option.

Fixed rate mortgages

These can be set up in various ways. Traditionally, the mortgage runs for 30 years with fixed monthly payments based on interest rates at the time the loan is initiated. Each payment covers a portion of the principal and interest. The early payments will be primarily for interest. Gradually, as the principal is repaid, the ratio will change until payments are primarily for principal.

The 15-year fixed rate mortgage is becoming more popular. It is generally offered at a lower interest rate than a 30-year mortgage. While it requires a higher monthly payment than a 30-year loan, you own your home in half the time. For instance, a $100,000, 30-year, 11 percent mortgage at a monthly payment of $952.32 has a pay-off cost of $342,836. The same 11 percent mortgage for 15 years carries a $1,136.60 monthly payment for a total of $204,587, a savings in interest of $138,249.

Adjustable rate mortgages (ARMs)

These are also referred to as variable rate mortgages (VRMs) or adjustable mortgage loans (AMLs).

ARMs have become increasingly popular with both borrowers and lenders as they offer borrowers a lower initial interest rate which moves up or down with the market. ARMs give the lender protection against holding mortgages that generate less income than the institution must pay out in interest on deposits. In an interest-sensitive market where rates change rapidly, most lenders prefer ARMs.

ARMs are more complex than fixed rate mortgages. Factors to consider are described below.

❏ *Discounts.* Lenders will offer discounted interest rates for an initial period, usually one year, as an inducement to accept an ARM. At the end of the specified period, the rate adjusts to the market level. If interest rates have increased substantially, then your rate can also.

❏ *Indexing.* Lenders adjust the interest rate periodically according to standard indexes such as U.S. Treasury bills. Some indexes have a slightly higher fluctuation rate than others. Ask to see a 5 or 10 year history of the index used by the lender to get an idea of the rate of change you may be facing.

❏ *Caps.* Every ARM should have a lifetime or aggregate cap. This is your protection against skyrocketing interest rates. For instance, if an ARM starts at 10 percent with a 6 percent cap, the interest rate cannot go higher than 16 percent during the life of the loan.

In addition, it is important to negotiate a periodic rate cap that controls the amount the interest payment can be increased at any one time. Usually, this will be set at 2 percent. A 10 percent ARM with a 2 percent yearly cap could only go to 12 percent the first year, 14 percent the second year, and so forth.

There are also payment caps that govern the dollar amount you can be required to pay. For instance, if the yearly rate were to increase to 15 percent, causing the payment to exceed the cap, the difference would be temporarily forgiven and added to the loan for payment later. This is negative amortization, which increases the amount of interest paid over the life of the loan. The result is a larger debt than originally negotiated.

❏ *Renegotiable and convertible loans.* These are other types of ARMs. With a renegotiable mortgage, the interest rate is set for a longer period of time than standard ARMs—perhaps 3 to 5 years. The mortgage is then renegotiated and "rolled over" into a new loan at the prevailing interest rate, either as a fixed rate mortgage or as another ARM. A convertible ARM may be converted to a fixed rate mortgage at the end of a stipulated adjustment period.

❏ *Balloon mortgages.* Balloon mortgages offer equal monthly payments of principal and interest with a single large final payment at the end of a specified number of years, such as 5 or 10. Some lenders will guarantee to refinance when the balloon is due, but will not commit to a specific interest rate.

The Consumer Handbook on Adjustable Rate Mortgages from the Consumer Information Center or *Fannie Mae's Consumer Guide to Adjustable Rate Mortgages* offer further information on ARMs.*

POINTS

When you start to negotiate a loan, the lender will specify an interest rate plus "x" points. One point equals one percent of the total loan. Points are levied by the lender to increase the yield on the loan to make it competitive with other loans in the marketplace. Thus, if you are quoted an interest rate "plus 2 points" on a $100,000 loan, you will pay a one-time fee of $2,000 in points at closing.

---

* *Consumer Handbook on Adjustable Rate Mortgages [420-T]*. Pueblo, CO: Consumer Information Center, 1984; *Fannie Mae's Consumer Guide to Adjustable Rate Mortgages*. Washington, D.C.: Federal National Mortgage Association, 1988.

When money is plentiful and the demand for mortgage loans is low, buyers can "comparison shop" among different lenders for the best overall loan package—rates, terms, points, and penalties (fees charged for paying off a loan before the final payment is due.)

ANNUAL PERCENTAGE RATE (APR)

Keep the APR in mind as you consider your mortgage options. This is the annual cost of credit over the life of the loan. The APR includes interest, service charges, points, loan fees, and mortgage insurance, plus sundry other charges, and is therefore higher than the base interest rate. It is the mortgage market equivalent of unit pricing on supermarket shelves, with a consistent standard for comparison shopping among mortgages with different terms and features. Your lender is required to tell you what the charges will be on the amount of money you expect to borrow.

With an adjustable rate mortgage, the APR is based on the estimated total value of the loan because you do not know how your interest rate may change in the future. Therefore, the APR will not reflect what you must actually pay. The lender will calculate the APR based on the assumption that current interest rates will remain in effect for the life of the loan. If you are thinking about an adjustable rate loan, be sure to factor in the APR.

APPLYING FOR A LOAN

When you go in search of a commitment for permanent financing, lenders will ask you for specific papers. You will need:

- ❑ Contract or binder on the land you have purchased, with proof of down-payment, a copy of the deed, and a property survey showing the position of the proposed house.
- ❑ Appraisal or contract of sale on your current home if you have one and are planning to sell it.
- ❑ Builder's construction contract or estimate for proposed house.
- ❑ Blueprints and specifications for proposed house.
- ❑ Appraisal based on blueprints of proposed construction.
- ❑ Personal financial history, including salary(s), other income, past and current creditors.
- ❑ An employment verification form from your employer.
- ❑ Tax returns for the past three years.

Other paperwork requirements vary from lender to lender. All lenders will ask you to fill out an application form.

Chapter Five

CUTTING COSTS

*Y*ou have the land, a set of spectacular plans, a wish list. But when the builder's bid comes in, you find you're over budget. What can you do? First of all, sit down with the architect and/or builder to study the plans. Are there any fundamental plan changes that will save you money? Most building materials come in 2- and 4-foot units. If the plans call for a 23-foot-long living room, you will be wasting a foot of materials. Adjusting the size to 22 or 24 feet saves both material and labor costs.

Keep in mind, however, that shaving a foot here and there hoping to save "x" dollars for every square foot you eliminate probably won't solve your budget problems. It is far more effective to downgrade costly features or postpone their installation to a later date (provided the delay won't be more difficult or expensive in the long run). On the builder's estimate, some items such as cabinets and appliances are listed as "allowances." This means that the estimate will cover (or "allow for") the stated cost of those items, but you are free to select other makes or models costing more or less. The builder will adjust the estimate accordingly.

COST-CUTTING OPTIONS

Consider the following cost-cutting options:

❏ Using standard plans and working directly with a builder instead of hiring an architect (as discussed in Chapter 2).

❏ A slab foundation or crawl space instead of a full basement (as discussed in Chapter 7).

❏ Vinyl, aluminum, or plywood siding instead of wood, brick, or stone.

❏ Asphalt shingles instead of cedar shakes for the roof.

❏ A carport instead of an enclosed garage.

- ❏ Flat rather than sloped or cathedral ceilings.
- ❏ Fewer windows, and some with fixed glass instead of windows that open and shut.
- ❏ Medium-grade wall-to-wall carpeting instead of hardwood floors.
- ❏ Vinyl floors and wall coverings instead of tile.
- ❏ Synthetic or cultured marble rather than quarry marble in the bathrooms.
- ❏ "Stock" kitchen cabinets instead of cabinets that are custom designed and built.
- ❏ Standard lighting fixtures, appliances, bathtubs and sinks for now (you can always upgrade later).

Crown moldings, chair rails, stained glass, built-ins—each of these special details will add to your costs. Decide which features are essential to your dream home now and which can be added later. Consult with your builder on other ways to save money.

Postponing

If you don't want to sacrifice details, consider leaving portions of the house unfinished for now. Can the game room, fireplace, and bath in the basement wait until another year? Or the children's attic bedroom/playroom suite? What about deferring the garage, a deck, or major landscaping? Builders contend that you may not save much by waiting, because the crew is already there working on the rest of the house; bringing workers back to finish the rooms or landscaping a year or two later can be costly. If cash flow is a problem now, however, postponing may be your best—or only—option.

Doing it yourself

How handy are you with a paintbrush? How patient? How much time do you have? Perhaps you can make arrangements to do the interior painting yourself. Or perhaps you will want to do the drywall or trimwork with the help of friends. This will require the builder's cooperation, building time into the schedule, and perhaps delaying move-in day. Many custom builders and lenders will not go along with this. They insist on a "turnkey" arrangement, where the builder completely finishes the home before "turning the keys" over to you. And lenders prefer to protect their investment in your home by requiring that professionals do the work.

If you perform some of the work yourself, will there be a delay in the construction loan draw and the conversion to permanent financing? If so, the schedule change must be arranged with both the builder and lender. Remember that carpeting can't be laid until the painting is finished. The final inspection by the lender is usually made after everything is completed. Consider carefully the delay your handiwork may cause.

Landscaping is another potential do-it-yourself project. However, most local governments require houses to have some form of landscaping before a certificate of occupancy can be issued. Discuss landscaping requirements with the local building department and the builder. (Landscaping is discussed further in Chapter 9.)

Acting as your own general contractor

How much can you save by acting as your own contractor? Estimates go up to 10 or 20 percent, but there are many angles to consider and potential savings can be eaten up by your mistakes.

One prime concern is financing. Finding a lender to finance the construction becomes significantly more difficult if you decide to run the show yourself. As noted above, lenders are leery of novices in this complicated business. They associate lack of experience with delays, shoddy construction, and even failure to complete a structure.

Subcontractors present another problem. They prefer working with someone whom they know and respect. They may like you, but if your inexperience causes them delays and callbacks, their schedules suffer. Moreover, you are a one-time customer to them. They prefer working with contractors who will give them steady business, so you may not get their speediest service.

But say you find reliable subs who agree to do the work. You can be sure they will charge you higher rates to compensate for all possible delays and problems, and because you represent one-time rather than repeat business. So you will undoubtedly pay more than an established builder would. And, regardless of the experience level of the subs, are you qualified to recognize whether the job is done right or not? Can you judge the quality of framing or wiring or roofing?

Suppliers may not want to extend credit to you or give you the volume discount they offer to steady builder customers. And relationships with local building departments, banks, and their inspectors are important. Who will get the fastest service? You can be sure the established builder will.

Time is another consideration. If the job is to be done right, you must be on the site every day—not for a few minutes or an hour, but for the better part of the day. Talk with friends who have acted as general contractor on the construction or remodeling of their houses. More than likely, their stories of the headaches, delays, budget overruns, lack of cooperation, and tremendous time commitment will dissuade you from doing the job yourself.

To those brave souls who still want to be their own general contractors— good luck.

Chapter Six

CONTRACTS AND INSURANCE

*A*s you go through the process of getting your dream house built, you will need contractual agreements and certain types of insurance to protect both you and the professionals involved in the design and construction of your home.

CONTRACTS

Explicit, well-thought-out contracts protect you and the supplier from misunderstandings and unnecessary expense. They also provide a basis for recourse if something does go wrong. It is best to seal all agreements with both a handshake and a signature.

What contracts will you need? This will depend on the location and complexity of the undertaking. No standard contract exists that is suitable for every situation, and what follows is by no means a comprehensive list of all provisions a contract should contain. It is advisable to have an attorney review any contract having to do with your new house.

Land contract

This should be prepared by an attorney, and should contain a description of the property, price and payment arrangements, title, and title insurance binder (the insurance company's commitment to issue a policy at closing).

The description of the property should be based on measurements by a licensed surveyor. It should include the location of property corners and locations of improvements, encroachments, and easements (discussed in

Chapter 3). The American Land Title Association (ALTA)* warns that any statement on a property survey such as "For Mortgage Loan Purposes Only" limits the survey's legal validity. To be legally binding, the survey must be signed and dated by a licensed surveyor in accordance with ALTA standards.

It is also important to have title insurance on the property. A lender's title insurance policy does not cover you; it only guarantees the lender recovery of its interest, which is the mortgage outstanding on the land. Title insurance is discussed in more detail below. You may also wish to add contingency clauses to the land contract. These are terms added to a contract that must be met in order for the contract to become binding. Such a clause might read: *The buyer's obligations under this agreement are contingent upon percolation test results and soil bore results satisfactory to buyer and buyer's engineer, on or before May 15, 19XX.* Receipt of building permits or securing loan commitments are other typical examples of contingencies.

Architect contract

The American Institute of Architects (AIA) has developed a series of standard contract forms, including one for residential design which can be used to suit the circumstances of the particular job.** It spells out what the architect will do, how much it will cost, and the method of payment. The contract covers preliminary plans, final blueprints, specifications, involvement in selection of the builder, and bids. It should also define the architect's responsibilities for monitoring construction, and any other supervisory responsibilities.

Note that AIA contracts are written to protect the architect's interests, so you will want an attorney to review and modify the contract you use as needed.

Builder contract

Because of variations in local codes and other factors, the National Association of Home Builders does not recommend a standard builder's contract, but has established general guidelines for builders to use in writing contracts.*** Local builders associations may have standard forms for use in

---

* American Land Title Association, 1828 L Street, N.W., Washington, D.C. 20036, 202/296-3671.

** To obtain copies of AIA contract documents, contact AIA Bookstore, 1735 New York Avenue, N.W., Washington, D.C. 20006, 202/626-7475.

*** *Builder's Guide to Contracts and Liability*, National Association of Home Builders, Washington, D.C., 1987, available from NAHB Bookstore, 15th and M Streets, N.W., Washington, D.C. 20005, 202/822-0463.

their areas. Some builders use *AIA Document A201—Builder Contract* as a guide. And many lenders can supply a standard builder contract. Whatever builder contract is used, it should include the provisions described below.

❑ Site location, the address, approximate dimensions of the house to be built, and a legal description of the site with sketch showing location of the house on the lot and clearance limits, if applicable.

❑ Detailed description of work to be performed. Plans and specifications should be included (plans and "specs" may be an addendum signed by both you and the builder).

❑ Construction schedule: when work is to begin and date of substantial completion. Most builders will refuse to include a penalty clause for failure to complete the job by the specified date. In fact, most contracts provide for extensions in the event of weather delays, unavailability of materials, and other factors beyond the builder's control. Suppose a roofer plans two days on your home, Thursday and Friday. It rains Monday, Tuesday and Wednesday of that week. The roofer must now make up the work scheduled elsewhere for Monday, Tuesday, and Wednesday before he can come to your site. This could delay roofing on your house.

❑ Differing site conditions clause. Unexpected site conditions such as ledge, bedrock, or an unexpectedly high water table could be discovered after the contract has been signed and work is in progress. Working around these abnormal site conditions can be very costly. You and the builder should agree on a procedure for handling any plan changes, schedule changes, and expenses incurred in the event differing site conditions are found.

❑ Fee and payment schedule. Amount of binder, due dates for payments, interest on late payments, and fee or percentage of profit on change orders (discussed below). The builder may require that evidence of your financing terms be included in the contract and produced prior to start of construction. Note that final payment to the builder is normally withheld until the house is completely finished, the walk-through has taken place and any problems corrected, and an occupancy permit from the building department has been received. At the time of final payment, the builder transfers the keys to you, the owner.

❑ Procedure for change orders. Alterations or additions to the plans or specs must be authorized in writing for both the builder's protection and yours. The contract should clearly define the change order procedure. Many changes can be performed easily; however, most will cost additional money. Each change order should specify what revisions are to be made, the cost, and any delays that will affect the completion date. The policy for changes due to builder error and the resulting costs must also be clearly stated.

What if you and the builder can't agree on who is responsible for a plan change? The contract should state that work will continue on the project and that you will pay for the work pending resolution of the dispute. The builder receives a set percentage of profit on the work if it is determined that he/she is not responsible for the error.

The contract should include a clause stating that all instructions and change orders will be given to the builder and only to the builder—not to subcontractors or crew members.

A good contract will also specify that the builder is not responsible for plan changes required by government authorities or particular conditions on the job that are unforeseeable, changes needed to correct or improve the design at the architect's authorization, or changes due to defects in the plans or specifications supplied by someone other than the builder.

❏ Your designated agent. If you are retaining the architect or a construction manager to monitor construction, the contract should clearly state that he/she is your designated agent; that the builder is to act upon the agent's instructions, revisions, and approvals; and that the builder is absolved of liability for any actions taken on the agent's say-so. If, on the other hand, you plan to work directly with the builder, the contract should cover this and establish liability (discussed below).

❏ Site improvements to be performed and who is responsible.

❏ Compliance with local codes and inspections.

❏ Who pays taxes, utility bills, and other fees during construction.

❏ Who owns the plans and specifications.

❏ Warranties. It is a good idea to inspect the warranty documents, the builder's walk-through or pre-occupancy procedures, and the service policy. These can be incorporated into the contract. Note that manufacturers' warranties on materials such as roofing, windows, and appliances do not cover installation. If a guaranteed window is installed improperly, the manufacturer is not liable. Who is? Clarify this. A reputable builder will correct any such defect and the warranty should so state. (Warranties are discussed further in Chapter 2.)

❏ Arbitration clause. It is wise to include a provision for arbitration in the event a dispute between you and the builder cannot be resolved. Arbitration is generally preferable to going to court. It costs less, and arbitrators are usually experts who may be more familiar with the technical aspects of a construction controversy than a jury would be. Arbitration is usually conducted by an impartial organization such as the

American Arbitration Association or the National Academy of Conciliators.* Some local builders associations also offer dispute resolution mechanisms.

The arbitration clause should specify notification procedures and time limits. All parties should understand that the arbitrators' decision is final and binding.

❑ Contract termination. The contract should set forth a procedure to be followed in the event the contract is terminated by either you or the builder, or due to an uncontrollable circumstance. For example, should you terminate the contract for any reason other than builder error, the contract may state that the builder will keep any money earned up to the date of termination, and/or that the builder may sue you for the value of the contract had the job been completed. If the builder terminates, the contract may state that you can seek a refund of money already paid to the builder. The contract should also provide for compensation or damages relating to attorney fees incurred as a result of terminating the contract. Bankruptcy by either party should also be addressed.

❑ Liability and insurance terms.

INSURANCE

Any construction project requires insurance. You and your builder are each responsible for providing certain types of insurance, which are discussed briefly below.

Liability insurance

You could be liable for an accident that occurs on your land, even though a person might be trespassing. Although your builder and crew are covered by workers compensation (discussed below), your liability insurance policy protects you against any freak claims resulting from their accidents. Consult with your insurance agent and an attorney to determine what level of coverage to get.

Title insurance

The title search covers recorded documents including deeds, judgments, taxes, assessments, and other matters. The title search cannot absolutely ensure that no problem exists, however. A deed may have been forged, for instance, or a lost heir may appear. In one case, property was bought from the heirs to a particular estate. Years later, another will was found leaving

---

* American Arbitration Association, 1730 Rhode Island Avenue, N.W., Washington, D.C. 20036, 202/296-8510; National Academy of Conciliators, 5530 Wisconsin Avenue, N.W., Suite 1250, Chevy Chase, MD 20815, 301/654-6515.

the property to different heirs and rendering the sale invalid. Title insurance protected the owner from loss by paying the full amount of the purchase price.

Two types of title insurance are available. Lenders' title insurance, which the lender usually requires you to provide, is issued at the time of closing and involves a one-time premium. It pays the lender's legal expenses for defending against a title claim and covers the amount of the mortgage in the event the lender loses the property. The level of coverage decreases as the loan is paid off. Lenders' title insurance does not compensate you, the buyer, for any legal expenses you may incur or the value of property you may lose. A separate owners' title insurance policy must be carried for that purpose.

An owners' policy insures that the title is as stated and protects you against a faulty title search. It usually contains an inflation rider that will increase the level of coverage as property values rise. Coverage remains in effect for as long as you or your heirs hold the property. While owners' insurance is not required, it is relatively inexpensive and the protection it affords is well worth the investment.

Owners' and lenders' title insurance can be secured simultaneously by the same title search and from the same company, lowering the overall cost to you.

Mortgage insurance

Your mortgage lender may require mortgage insurance, depending on the amount of equity you have in the home. Usually, if you have less than 80 percent equity in the home, you must carry mortgage insurance. This protects the lender in case of default.

Many people, particularly with young children, carry mortgage life insurance which pays off the mortgage in case of disability or death. This is a personal insurance policy and has nothing to do with the lender's mortgage insurance. Probably the best buy is term insurance, which is a form of life insurance protecting the policy holder for a specified period of time (or "term"). Term insurance can cover the full amount of the mortgage and decreases in value and premiums as the principal is paid off.

Workers compensation

Workers must be covered by workers compensation insurance during construction to protect them against injury and death in the course of employment. It is the builder's responsibility to see that all workers are covered. The builder normally carries workers compensation for his/her crews, and subcontractors cover their own people. The builder is required to maintain current workers compensation certificates for all subcontractors. If you do any of the subcontracting yourself, it is your responsibility to see that the subcontractors carry the proper insurance. Your contract with the subcontractor should clarify this.

Builders' risk insurance

Insurance on construction is essential to protect against vandalism, theft, structural collapse, and other building site hazards. The builder usually carries this during construction (although it is a negotiable contract item). Be sure you know what the insurance covers, and ask for a copy of both the certificate of insurance and the policy. The builder may want you, the owner, to be named on the policy, which could relieve you of the need to carry separate liability insurance (if you are satisfied with the scope and level of coverage). Note that if you already own the lot on which the house is to be built and your money is paying for construction, you may be the one to secure risk coverage. Again, you and the builder can negotiate this point. Your policy should include coverage for the builder and subcontractors. Check with your insurance agent about your protection options.

Homeowners' insurance

Once the house is completed, you will need homeowners' insurance covering damage, fire, theft, and liability. At closing, your lender will require you to have a policy with one year's premium prepaid in full. Policies vary. The best—and most costly—are the all-risk policies which cover everything not specifically listed as an exclusion. You can also obtain broad coverage that covers a variety of specified perils. These will not cover earthquakes, floods, and various other disasters. However, you can purchase earthquake insurance in a separate policy or as a rider to the basic homeowners' policy. Your lender will require you to provide flood insurance if the property is located in an area that has been designated a flood hazard area by the Federal Insurance Administration (FIA). Chapter 3 provides additional information on flood hazard.

Chapter Seven

HOW A HOUSE IS BUILT

*F*rom foundation to finish, human hands craft custom homes and often can't be hurried. That is what makes each house distinctive and unique. Yet many homeowners are baffled by what appear to be unnecessary delays, days when the builder and crews simply don't work.

The house construction process involves a complex network of people and activities. You, the owner, profit from the custom builder's skill in planning and scheduling this network. You benefit from better subcontractors and better quality work, jobs completed more quickly, and fewer headaches.

It is in the builder's interest, too, to plan and schedule efficiently. Builders who have top-notch relationships with subcontractors get better prices from their subs and first call on their time. Suppliers and lenders prefer to work with seasoned professionals who know the construction process inside out.

But even the best scheduled job will encounter unforeseen delays. It is easier to understand why a custom house can require 6 or more months to build when you realize how many steps are involved in building a 2,500-square-foot house. Understanding the various stages of house construction and their complexity can make life simpler for everyone. This chapter walks you, the homeowner-to-be, through the process of building a house.

BUILDING CODES

Construction in most areas of the country is regulated at the local level by building codes (although some rural areas have no building code regulations). These codes govern construction; plumbing, electrical, and mechanical systems; and fire safety. While a few municipalities (mostly major cities) write their own codes, most state, county, or local jurisdictions adopt model codes that have been prepared by 4 major model code service organizations: the Building Officials and Code Administrators International (BOCA); the International Conference of Building Officials (ICBO); the Southern Building Code Congress International (SBCCI); and the Council of American Building Officials (CABO), which is a federation of the 3 preceding groups. Each group's codes influence a different region of the country: BOCA in the northeast and northern midwest; ICBO in the midwest and west; and SBCCI in the southeast and most of Texas.

Health codes, which are established and maintained at the county or municipal level in most parts of the country, govern wells and septic systems. Public water and sewer are usually controlled by county or municipal building or engineering departments.

You will want to find out whether any code requirements will make building on the site you have chosen overly expensive. Talk to the builder or architect about this or check with the local building department. Your architect and builder should be well versed in the applicable codes. Their compliance with the codes is ensured through the local jurisdiction's inspection process. Note that inspections do not evaluate quality of construction—only code compliance.

PERMITS AND INSPECTIONS

The builder will generally be responsible for obtaining permits and inspections throughout the construction process. The procedures differ in each locale, so check with the builder or your local building and health departments. Permits cost money, so budget for them if you are handling this aspect yourself. Before ground is broken, you will generally need preliminary building plan approval from the building department, from the zoning or planning commission, and from the health department if you are planning a septic system and/or well. Depending on the jurisdiction, you may also need a permit from the local environmental authority.

After permits have been issued and construction begins, inspections are required at specified stages of completion. The builder will inform the appropriate department when the house is ready for inspection, and the inspector will leave a "passed" (or "failed") notice on the house. Only when the work has passed inspection can building continue.

While inspection requirements vary from community to community, the following inspections are typical:

❑ Building or engineering department inspects municipal water service and sewer connections (unless well and septic system are to be installed).

- [ ] Building department inspects footings, open trenches and/or formwork before concrete is poured. If steel reinforcement is used, it is inspected at the same time. Footing depth and soil conditions are checked to ensure that the footings will provide adequate support for the structure above.
- [ ] Building department inspects foundation prior to waterproofing and backfilling.
- [ ] Health department inspects well and septic system.
- [ ] Building department inspects roughed-in framing, plumbing, electrical, and mechanical systems.
- [ ] Building department performs final inspection to check plumbing, electrical, and mechanical systems, interior and exterior finish, and landscaping. If all is in order, a certificate of occupancy is issued. (In some municipalities, the local board of fire underwriters must inspect the electrical installation before a certificate of occupancy can be issued.)

DELAYS

Assuming that all permits are in order and building has begun, what can cause delays?

Deliveries are one cause. A truck breaks down, a factory goes on strike, a shipment of imported tile has to wait for customs clearance. Subcontractors are another cause of delays due to rain, illness, or problems on a prior job. Change orders, slow decisions, or unclear instructions from the architect or the customer can cause a back-up that affects the entire job.

Inspections are another culprit. The building inspector may take several days to get to the site, no matter what kind of notice is given. Yet construction cannot proceed without the inspection. Most banks also base their draw (or construction loan payment) schedule on inspections. Delays caused by waiting for inspections (and reinspections if corrections are needed) can throw off the bank schedule.

THE CONSTRUCTION PROCESS

To better understand how a house is built, let's examine the construction process from the ground up. Figure 3 shows a sample construction schedule for completion of a 2,500-square-foot house from groundbreaking to final inspection. The amount of time required to complete each phase of construction will vary depending on the complexity of the design, availability of materials, weather, and other factors beyond the builder's control and yours.

3. Sample construction schedule.

CONSTRUCTION PHASE	MONTH 1						MONTH 2					
Clearing, stump removal	■											
Grading, driveway	■											
Well		■										
Basement excavation		■										
Temporary utilities			■									
Foundation footings				■								
INSPECTION: footings					■							
Foundation walls						■						
INSPECTION: foundation							■					
Septic system installation							■					
INSPECTION: well, septic system								■				
Backfill								■				
Framing, sheathing, roofing									■			
Rough-in of plumbing, electrical, mechanical systems												
INSPECTION: framing, rough-in												
Exterior finish												
Basement floor												
Insulation												
Drywall												
Hardwood floors												
Interior paint, wallpaper, paneling												
Vinyl flooring												
Trim, cabinets, bathroom fixtures												
Appliances, carpet, tile, light fixtures												
Final grading, driveway, walkways, deck												
Landscaping												
INSPECTION: final												

MONTH 3	MONTH 4	MONTH 5	MONTH 6	MONTH 7

Stakeout

The first step is for you and the builder to do a preliminary "stakeout" of the house and driveway. The builder uses the plans to measure out the house dimensions on the lot, indicating each corner with a stake. He/she will make sure to maintain correct distances from lot lines in keeping with local zoning requirements. Now is the time for you to designate the trees you want saved and the views you want to capture.

Site preparation

The next step is clearing and grading the site. This includes removing trees, changing the slope of the land if needed, leveling, and installing a solid gravel driveway or road to provide access for the heavy trucks and equipment that will be serving the site throughout construction (Figure 4). Final stakeout is done, then the well is dug. In the unlikely event that the well driller has difficulty finding water, it is better to decide on the well location before the basement is dug. Water will be needed for the construction work anyway. Note that it can take up to 3 months for a well driller to obtain a permit for the well and dig it, so be sure to plan ahead.

4. Site preparation.

Excavation

Once the well is in place, the basement hole is dug (or "excavated"). The hole will be larger than the actual house dimensions to allow for working room and the installation of drains. The bottom is leveled off for the slab.

Excavation for utilities may be done at this time: trenches for hook-ups to city water and sewer or the septic tank, and gas and electric line trenches. Depending on local regulations and your sense of aesthetics, electric lines may be brought in by overhead cable. Electrical hook-ups should be made so that power tools can be used. These often are temporary hook-ups that are dismantled after the home's permanent electrical system is installed. Now is also the time to arrange with the telephone company for telephone lines if you will require unusually long lines or want a phone onsite during construction.

Footings

Next come the concrete footings. These are the supports for the foundation walls. They extend beyond the walls to spread the weight of the house over a greater area of earth, giving the structure more stability (Figure 5). Footings are reinforced with steel rods over trenches and soft soil for extra strength. (If your house will have a treated wood foundation, discussed below, gravel is used to spread the weight of the house instead of concrete footings.) One reason many very old houses sag is their lack of footings. Local codes usually govern the size of footings based on soil composition and the weight of the building.

Depending on local building department requirements, a footing inspection may be required at this point.

Foundation

The foundation walls come next. These can be built from concrete block, poured concrete, or wood treated with a preservative to resist insects and rot (Figure 6). If concrete block is used and the presence of radon is suspected, 6-mil polyethylene film is placed over the concrete block walls and capped with a solid course of block.

The foundation must then be waterproofed with a tarlike bituminous compound or, in poorly draining soils, membrane waterproofing. Drains or drain pipes are laid in gravel or crushed stone around the base of the foundation. They are sloped to carry water away from the foundation to a storm drain, to the ground surface, or to a sump pump located below the basement floor.

What if your dream house doesn't have a basement? Many new homes are being built over crawl spaces or on concrete slabs. A crawl space is an unfinished area between the ground surface and the first floor, usually just big enough to "crawl" through for maintenance and repairs. Crawl spaces cost less than full basements because they only require excavation and grading for footings and walls. Crawl spaces should be protected from ground moisture and radon with a layer of 6-mil polyethylene and foundation vents. To prevent decay and termite damage, untreated wood should not come in direct contact with the soil.

Concrete floor slabs poured directly on prepared soil (called "slab on grade") are becoming more common throughout the country. Construction techniques vary according to local building practice. For moisture protection, concrete slabs should be built on well-drained coarse sand or 4 inches of gravel or crushed stone topped with 6-mil polyethylene. Where radon is

5. Diagram of house under construction.

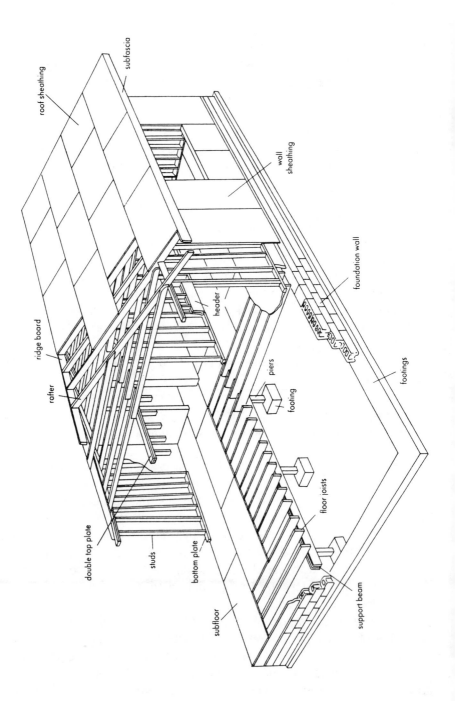

suspected, all cracks or holes in the slab should be sealed with durable, flexible caulking.

Crawl spaces and slab-on-grade foundations require less time to install than a full basement; however, allow for delays due to scheduling and weather. Installing a heating system in the slab can require additional time as well.

A building department inspection generally takes place upon completion of the foundation.

Septic system

At this stage the septic system is usually installed. The builder must obtain a permit from the county or local health department once a percolation test has been performed on your lot. (Depending on the jurisdiction, a health department representative or an independent contractor such as a soils engineer performs the "perc" test. The test determines the soil's water absorption rate and the best location for the septic system.)

A trench is dug from the place where waste will leave the house to an area of the lot where the health department has determined the septic field should be located. A septic tank is placed in the trench between the house and the septic field to collect and process waste. Once solids have settled to the bottom of the septic tank, a pipe will carry the liquid waste to the septic field, where it is gradually absorbed by the soil and purified.

A health department inspector will inspect the septic system to verify that it has been installed properly before the trench is closed.

The earth is replaced in the trench around the foundation and the septic trench; this is called backfilling. In most parts of the country, termites can be a problem, so the soil around a foundation should be treated with an approved pesticide.

Framing

Framing provides a skeleton or internal structure for the house, making it easier for you to visualize your new home. A wood or steel center beam is laid into the top of the foundation walls and supported between the walls by steel columns or piers. Sill plates are bolted onto the foundation walls, and floor joists—a series of horizontal parallel beams that support floors and ceilings—are nailed to the sill and center beam (Figure 5). Next the plywood, flakeboard, or oriented strand board subflooring is put on. This is the base on which floor finish (hardwood, carpet, vinyl, marble, or tile) is laid. Sometimes smooth sheets of "underlayment" are used between the subflooring and finish floor materials. Next come the studs—upright pieces of lumber or metal that will support the walls (Figure 7). Doors and windows are framed.

Roof framing is next, which involves the placement and fastening together of all the support members in a roof (Figure 5). The roof is sheathed (commonly with plywood, flakeboard, oriented strand board, or 1-inch lumber) to distribute wind and snow loads and to provide a solid base for fastening the roofing material. Roofing paper underlayment is usually applied to the sheathing, metal flashing is installed along "valleys" in the roof, and a metal drip edge is applied to the eaves to channel water off the roof. Now the roof is ready for its final layer. The choice of roofing material depends on budget, local codes, the design of the house, and individual preferences. Options include wood or asphalt shingles and shakes, tile, slate, or sheet metals such as aluminum, copper, and tin (Figure 8).

Now that you can walk through every room, you feel sure that your new house will be finished soon. It won't. The most time-consuming work lies ahead.

Exterior wall sheathing (plywood, flakeboard, oriented strand board, or other material) is nailed to the studs. It seals out air and helps to reinforce the studs.

Windows and doors are installed next, and the house can be locked up. (Note: bathtubs and one-piece bathtub/shower units won't fit through most finished doorways. These must be installed before the sheathing, windows, and doors go up.) You are now "under roof" and ready for the many components that will make this skeleton a home.

Rough-in of plumbing, electrical, and mechanical systems

Now a whole new set of trades moves into the house. Plumbers install piping for water circulation and sewage. Heating and air conditioning systems are installed by subcontractors who specialize in those trades (Figure 9).

Electricians lay cable and install the various boxes and switches that will allow your house to function properly. They also install wiring for the electrical circuits, thermostat, doorbells, intercom, telephone, cable television, and security system (Figure 10). Do some advance room use and furniture

7. Studs used in house framing.

8. Roofing process: sheathing, underlayment, shingles.

9. Installation of heating/air conditioning ducts.

10. Rough-in of electrical wiring.

planning to assist the electrician in locating these items conveniently for your future use.

The house is now a maze of wires winding through studs, shiny ducts for heating and cooling, and plumbing pipes of all kinds. This is the time for the next building inspection. Because roughed-in elements will be hidden behind walls and under flooring, the inspector must determine that each is installed according to code and that the framing has been built properly.

Exterior finish

Exterior finish can be started while the interior rough-in is taking place. The choice of exterior finish material depends on your budget, personal preferences, and regional style. Some builders wrap a woven air- and water-resistant polyethylene air barrier around the house before the exterior finish is applied; others use treated paper sheathing to keep out air and moisture. If stone or brick is being used to finish the house, the foundation will have been planned with a ledge to support it. Wood, aluminum, or vinyl siding and trim can be used, or stucco, and aluminum or vinyl gutters installed (Figures 11, 12). Chimneys and fireplaces can be finished during this time.

11. Brick finish being applied over sheathing paper.

12. Siding being applied over sheathing paper.

Interior finish

Now that the piping and electrical rough-ins are finished and inspected, the basement floor can be poured. Concrete goes over a 4-inch layer of coarse gravel and a 6-mil polyethylene vapor barrier. Be sure a sump or drain system of some kind has been specified.

Where radon gas is a possibility, your builder can use certain basic construction methods to reduce radon entry into the house. The Environmental Protection Agency (EPA) publication, *Radon Reduction in New Construction (An Interim Guide)* is a good source of information on construction techniques.*

Next comes insulation, which most commonly is made of mineral, glass, or cellulose fibers; or plastic foam. Insulation is placed in floors or foundation walls, above-ground walls, and in the attic—all the places through which air is likely to seep. It is covered with a vapor barrier (paper or plastic sheets treated to retard the flow of moisture into the insulation). The insulation and vapor barrier are covered with sheets of drywall (also called gypsum wallboard or sheetrock), which form the interior walls and ceilings of most houses.

The sections of drywall must be taped at the seams, which are spackled (or coated) with a joint compound, dried and sanded smooth to prepare them for painting or wallpapering (Figure 13). This job requires meticulous attention to prevent waves and visible joints.

Hardwood floors are installed next (although sanding and varnishing are not done until the very end). Rooms are painted or wallpapered, and paneling goes up. Vinyl flooring is laid, and interior doors, kitchen cabinets, the bathroom vanity and commode go in over the vinyl flooring. (If tile flooring or carpeting is used, it is laid after the cabinets and fixtures have been installed to minimize wear and tear.)

Medicine cabinets and bathroom hardware are installed, and moldings are applied around windows, doors, and cornices. Baseboards and toe molding go in last, especially when wood flooring is used. Then appliances, tile floors and carpet, light fixtures and light switch covers are installed. Hardwood floors are sanded and varnished. This can take a week or more, as the floor must dry between each application of finish.

Once the interior finish is complete, the house receives a final thorough cleaning.

---

* *Radon Reduction in New Construction: An Interim Guide.* Washington, D.C.: U.S. Environmental Protection Agency, 1987. This publication is available free of charge from your state health department or environmental office, or the EPA regional office.

13. Taped and spackled drywall.

After all the delivery trucks have come and gone and the trash dumpsters have been hauled away, the time has finally come for final grading, installing walkways, decks, and outside lighting, surfacing the driveway and landscaping. The building inspector gives the house a final inspection and issues a certificate of occupancy.

MAKING CHANGES

One advantage to having a custom home built is flexibility in making changes to the original plans. Changes can be both costly and time-consuming, however, so how can they be controlled?

One mortgage banker tells the woeful tale of a new home where the owners made $75,000 worth of changes during construction and then couldn't come up with the money to pay for them. The bank was saddled with a house to sell, and the owners lost their dream house. Who was at fault? The builder failed to make the owner sign for each and every change and did not bill for those changes monthly. But the owner was also at fault for being indecisive and unrealistic. And in the end, everybody lost.

Changes can be expensive because of the extra architectural, supervisory, labor, and purchasing costs involved. Changes also delay construction loan draws, which can create financing problems. Take time to study the preliminary plans carefully, and make sure they meet your needs before the final plans are drawn. Only changes that are absolutely essential should be made to the final plans. When you are completely satisfied, the builder can break ground.

But what if you realize a change must be made during construction? If the builder says your request to move a door or change a window is feasible, ask how much it will cost and how long it will delay the job. Insist on getting all changes written down, signed and dated, even if your builder doesn't require it.

Sometimes changes are unavoidable. Then the architect, the builder, and you should sit down and talk. The architect will not be responsible for paying for the changes, but should be willing to revise the plans without charge. The builder will not absorb additional costs incurred because of the changes, so you will be financially responsible. Be sure there are no misunderstandings in anyone's mind about this.

If your architect is monitoring construction, part of the responsibility is to ensure that plans are followed, and that any changes are made for good reason (such as an unusual delay in deliveries). Unless you give the architect specific authority to make changes, all changes must be approved by you. The same goes for a construction manager if you use one.

Use the following checklist to be sure you use the correct procedure for making changes.

- ☐ Get in writing the costs of material and labor plus the amount of delay each change will cause, and sign and date it.
- ☐ Never give instructions to anyone but the builder. The architect must also follow this rule. Let the builder tell the crews and subs.
- ☐ Don't argue with the builder in public. It will weaken his/her authority with crews and subs and, consequently, hurt construction of your house.
- ☐ Don't delay decisions. Delays can hold up work, destroy schedules, and cost you in the long run. Approve or reject work and make decisions quickly.

Chapter Eight

REMODELING YOUR HOME

*S*uppose your current house is situated on a lot you love. Property values in your area are appreciating nicely, you are satisfied with schools and community amenities, and your daily commute to work is short and easy. There is only one problem: you've outgrown your home.

One solution is to have your house remodeled into the home of your dreams. Remodeling allows you to combine the best features of your current home with a whole wish list of new design ideas. Depending on your needs, wants, and budget, remodeling can be as simple as converting that underused kitchen pantry into a much-needed powder room, or as complex as adding a multi-story wing to your existing house.

Many of the steps involved in remodeling are the same as those discussed earlier for new home construction. These include:

❏ Careful planning—creating a design file of ideas for your dream home and putting together a wish list.

❏ Working with professionals—your remodeling contractor and perhaps an architect and/or an interior designer—to translate your ideas into living space.

❏ Budgeting for the remodeling job, and knowing how and where to get financing if needed.

❏ Using contracts and insurance to protect both you and the remodeling team throughout the project.

❏ Having a basic understanding of the construction process.

This chapter focuses on those considerations that are unique to remodeling.

77

PLANNING THE REMODELING JOB

Start by identifying what you want to accomplish by remodeling your house. Do you need more space? A new room configuration? Or simply an updated look? As you think about your needs and wants, evaluate the constraints and opportunities that currently exist in your house.

If you are feeling cramped for space, altering the existing room layout may solve the problem. This could be accomplished by moving or removing a wall (as long as it is not a "load bearing" wall that supports the structure of the house) widening a doorway, or opening up a little-used closet to extend a room. Removing a wall between the kitchen and dining room can create a single large "great room," which gives a feeling of more space and is well suited to informal meal preparation, dining, and entertaining. Or simply widen the passageway between the two rooms to open up the space visually.

Or do you want an addition built onto the existing house? Building up or out requires breaking through the exterior walls and perhaps the roof. This is generally more disruptive to household routine, and more costly—but it offers more design possibilities than working with existing space. Note that zoning regulations may affect your plans, so check with the local planning department early in the planning process.

As with the construction of a new home, you will want to create a design file of ideas gathered from books and magazines, the homes of friends, and your own wish list of features that are important to you. And as with new construction, be sure that you and your spouse agree on what you want done before calling in a remodeling team.

WORKING WITH THE REMODELING TEAM

Once your ideas are down on paper, you are ready to work with the people who will make your dreams a reality. Whether you use an architect to design your remodeling job largely depends on the project's complexity, your finances, and the services the remodeling company can provide.

Many remodeling contractors have the experience and technical skills required to transform your ideas into finished rooms. Some remodelers have designers on staff to help you formulate your plans, as do the "design/build" firms discussed in Chapter 2. Many people find it more convenient to work with a design/build company than to hire a separate architect, for the following reasons:

❑ The customer has to deal with only one company for both design and construction, which can be convenient and cost-effective.

❑ The designer understands what type of work the remodeler can do based on the customer's needs and budget, and plans accordingly.

❏ The designer and contractor are accustomed to working together, and communication between them is often more frequent and more productive than communication between two separate companies would be.

However, if you seek extensive and highly innovative design guidance, you may wish to start the remodeling process with an architect's plans. Chapter 2 provides further information on selecting and working with an architect.

Selecting a remodeling contractor is much the same as finding the right builder to build a new house. Your local builders association, affiliated with the National Association of Home Builders, can provide you with a list of members who offer remodeling services. Word-of-mouth referrals from friends and colleagues are another excellent source. Keep in mind that the lowest bid is not always the best bid for your project. The remodeling company's reputation and compatibility with you and your needs are equally important.

FINANCING THE PROJECT

Depending on your financial resources and the complexity of the remodeling to be performed, you may need to obtain a loan to cover the costs you incur. Financing options include—

❏ A personal loan from family or friends.
❏ A signature loan.
❏ A home equity loan (also called a home equity line of credit).
❏ A second mortgage.
❏ Refinancing your home.
❏ Borrowing against insurance, investments, and corporate savings plans.

Personal loan

Borrowing money from family or friends is the least complicated way of obtaining a loan, at least in terms of paperwork. If you do borrow from friends or family, be sure that both parties agree in advance to the terms of the loan—the amount being borrowed, how soon you will repay the loan, and at what interest rate, if any. It is advisable to put the agreement in writing, signed and dated by both parties. Nothing can ruin a friendship or family ties more quickly than a dispute over money.

Signature loan

If the amount of money you need to borrow for the remodeling project is relatively small—under $10,000 or so—you may want to obtain it through a signature loan from your bank. This type of loan is based on your ability to pay, and is guaranteed by your signature. Your bank decides how much it can lend you based on your income, your credit history, and your record with that bank. If you fail to pay back the loan, the bank may use your personal assets for the money owed. Signature loans usually carry a fairly high interest rate, but they generally require no collateral and no closing costs.

Signature loans are available from commercial banks, savings and loans, credit unions, and consumer finance companies.

Home equity loan

Also called a home equity line of credit, this method of financing involves establishing a line of credit with a bank, using the value of your house and any other assets you may have to secure the loan. You draw funds as you need them up to a predetermined amount. Home equity loans are generally available through commercial banks and savings and loans. They have become very popular for financing remodeling projects because the interest is tax-deductible.

If you opt for a home equity loan, be sure you understand the terms. Many banks offer a low introductory interest rate that jumps to market rate after a specified length of time. This increase can strain your budget unless you are prepared for it. Depending on the amount of equity you have in your home, you may be required to pay closing costs such as those described in Chapter 10. Your budget should allow for these costs as well. Ask your lender for details on obtaining a home equity line of credit.

Second mortgage

Obtaining a second mortgage on your home allows you to finance remodeling by using the value of the home as collateral on the loan. Second mortgages most frequently have 10- or 20-year terms, although terms as short as 2 years or as long as 30 are also available. The lender from whom you obtained the first mortgage has first call on any funds generated by the sale of the property in the event of foreclosure.

Because fees on second mortgages can be high, most lenders will advise against obtaining a second mortgage if you can borrow what you need through a signature loan, from a credit union, or from other sources.

Refinancing your home

According to a common rule of thumb, you may want to refinance your home if current interest rates are at least 2 percentage points below the mortgage interest rate you now have. This may enable you to increase your mortgage loan by the amount of money your remodeling will cost. A lender will usually refinance up to 75-80 percent of your home's current market value (up to 95 percent if you get mortgage insurance). Additional points, the lender's refinancing fee, and other expenses may make this option too costly. Discuss refinancing with your lender and compare the costs with those of a home equity loan or a second mortgage.

Other financing options

If your life insurance policy has a high enough cash value, you may be able to borrow against it to pay for the remodeling. Read the policy document carefully to understand the terms of this option, and discuss the arrangement with your insurance agent. Note that if you should die before the loan is repaid, the proceeds paid to your beneficiary will be reduced by the amount owed.

You may also borrow against your investments, using the current market value of stocks and bonds as collateral for the loan. If the market value of your investments should drop significantly, your lender may ask you to repay a portion of the loan right away.

If you are fully vested in a savings or profit sharing plan offered through your employer, you may be able to borrow against these investments to pay for the remodeling. Ask the plan administrator whether your company offers this option and what the terms are. Note that if you should leave the company before you have fully repaid the loan, the amount still owed will be deducted from the total you receive from the plan.

CONTRACTS AND INSURANCE
Contracts

As with new construction, you and your remodeling contractor will want a written contract that spells out:

❏ The contractor's responsibilities in detail.

❏ Work to be performed, including plans and specifications for materials to be used (quality, quantity, weight, color, size, model or brand name where applicable).

❏ Approximate start and completion dates for the project.

❏ Financial terms (including total price of the job, payment schedule, terms of cancellation penalty where applicable).

❏ Any special conditions and requests (such as clean-up and storage requirements, instructions for saving certain materials removed during the remodeling, policies regarding children and pets).

On any home improvement job, you should expect to make a downpayment that is approximately one-third of the total contract price. Except for the downpayment, you should not make payments for work that has not yet been completed. Instead, schedule payments at weekly or monthly intervals, or after completion of each phase of the project.

When you sign a home improvement contract with the remodeling contractor or the contractor's representative, you usually have 3 working days to change your mind and cancel the contract. The contractor must inform you of your cancellation rights both orally and in writing, and must provide you with the forms to use for cancellation. If you need emergency repairs, you can waive your 3-day cancellation rights. This waiver is important because remodeling contractors generally wait until the 3-day period has passed before starting work.

It is strongly recommended that you work with a remodeling contractor who offers a warranty. Read the warranty document carefully and make sure you understand the terms before signing the contract.

Chapter 6 provides additional information on contracts.

Insurance

Insurance requirements for a remodeling job are virtually the same as for new construction. The following types of insurance are required; they are discussed in further detail in Chapter 6.

- ❏ Liability insurance (provided by you).
- ❏ Workers compensation (provided by the remodeler and subcontractors performing the work).
- ❏ Builders' risk insurance (provided by the remodeler).

THE CONSTRUCTION PROCESS

House construction is a logical step-by-step process, as Chapter 7 describes. Remodeling follows basically the same sequence as new construction. Building and health codes must be followed and permits obtained. Inspections are required throughout the remodeling process, just as they are for new construction.

Depending on the complexity of the remodeling to be performed, your remodeling contractor will start with a foundation and move through framing, rough-in, exterior finish, and interior finish.

The length of time required to complete a remodeling project also depends on the project's complexity. Converting a pantry into a powder room could take 2 weeks, while adding a second story to the house could require up to a year. Also remember that a remodeling contractor is subject to the same delays as a builder of new houses would be—delays due to weather, material shortages, shipping problems, and so on.

LIFE DURING REMODELING

Your contractor and crew will make every effort to minimize disruption in your home during the remodeling. Nonetheless, some intrusion on your household's daily routine is inevitable. At the very least, unfamiliar faces will be in and around your home for days, weeks, or months. Construction materials and debris will be stored on your property. Your telephone and perhaps your bathroom will be used by strangers.

If the remodeling is extensive—new rooms or a second story added, the kitchen redone—you can expect far greater inconvenience. An exterior wall may be ripped away, exposing your household to the elements. Or you may be cooking on a hot plate and doing dishes in the bathroom sink for a couple of weeks.

Careful planning and communication among family members and the remodeling crew will make life during remodeling easier for everyone involved. Before you consult a remodeling contractor, you and your family should sit down and plan strategies for handling household disruptions. Depending on the type of remodeling to be performed and your household composition, strategies could include—

❑ Reworking schedules for family bathroom use if one bath is being remodeled.

❑ Scheduling remodeling for summer months while the children are at camp and the weather is warm.

❑ Moving out of the house for a period of time if the remodeling will affect large portions of the house.

Of course, each remodeling job is unique, designed to satisfy the dreams and requirements of you, the customer. The success of your home's transformation depends largely on good communication between you and the person or team who will be remodeling your home. If all parties understand one another at the project's outset, life during remodeling will be easier and the end product will be successful.

Chapter Nine

LANDSCAPING

*M*ost local jurisdictions require that some type of landscaping be in place before they will issue a certificate of occupancy. Many lenders will also want to see some landscaping before they release the final construction draw. This "landscaping" can be as minimal as grass seed strewn over the graded lot.

Grass alone will not do justice to your new home, however. A comprehensive landscape plan can make the home seem larger and more private. It defines areas of use throughout the lot: walkways, gardens, a play yard, a patio, pool, or tennis court. Good landscaping can also create vistas for each season: flowering trees in springtime, brilliant flowers in summer, a specimen tree whose foliage and shape delight the eye in every season. Even the smallest garden in the smallest yard can enhance a home.

Many builder contracts only cover landscaping through final grading after the house is completed, with perhaps the installation of a couple of shrubs. Soil preparation and plantings are usually separate items to be negotiated with the builder or a landscaping professional. Your budget should provide for landscaping. If you pride yourself on your green thumb, you may want to do some of the work yourself. However, professionals who plant shrubs and trees usually guarantee their greenery for a year. The plantings you do yourself have no such guarantee.

TYPES OF LANDSCAPING SPECIALISTS

If you decide to use a landscaping specialist, several levels and types of expertise are available to you. These are described below.

Landscape architects

Landscape architects can advise you on design and siting as well as grading, irrigation, drainage, lighting, patio design, and outdoor structures such as pools, tennis courts, and gazebos. While your architect or builder can handle siting the house on the lot, you may want to utilize the landscape architect's know-how at this important stage.

Landscape architecture is a licensed profession in 40 states. Licensing involves a degree in landscape architecture from an accredited program plus successful completion of the Uniform National Exam (UNE) administered by the Council of Landscape Architectural Registration Boards. Professionals may also qualify for membership in the American Society of Landscape Architects (ASLA) based on a combination of education and experience.

Landscape architects generally work on an hourly retainer basis. Fees range from $48 to over $100 an hour depending on the individual and the location. Some landscape architects also base their fees on a percentage of the total landscaping project cost.

Landscape designers

While all landscape architects are designers, not all designers are trained as landscape architects. Landscape designers generally do not handle overall concept, site engineering, outdoor structures, or lighting. They do handle plantings, and can design a plan that specifies type and location of plants. Many nurseries and garden centers have landscape designers on staff; some may also employ a landscape architect.

Landscape designer fees are usually based on a percentage of the total landscaping project cost.

Landscape contractors

Contractors work with landscape architects on the installation and maintenance of both exterior and interior plantings, and can advise you on the best plant choices for specific sites and conditions. Some contracting firms perform grading, irrigation, swimming pool installation, and other outdoor work; some also have design capability. Landscape contractors generally charge a flat fee for the project, plus an additional charge for maintenance if provided.

The Associated Landscape Contractors of America (ALCA)* requires its members to have a minimum of 3 years of landscaping experience.

Nurseries

Nurseries are probably the most familiar players in the landscaping field; you probably have bought plants from one in your neighborhood. Many

---

* Associated Landscape Contractors of America, 405 North Washington Street, Suite 104, Falls Church, VA 22046, 703/241-4004

nurseries have designers on staff, and some will act as contractors for grading, irrigation, and other ground work as well.

Nurseries specialize in plants. They offer a wide variety of trees, shrubs, bedding plants, and seeds; some nurseries also sell topsoil and sod. Most nurseries will install the plantings they sell and maintain them for a fee; many guarantee the plants they sell and install for one year.

SELECTING THE LANDSCAPER

Whether you retain a landscape architect or decide to work with the designer at your favorite nursery, homework will pay off. How do you go about finding a qualified landscaper? Ask friends and colleagues, your builder or architect, or the owners of a home whose landscaping you've long admired. Check with the local chapter of the American Society of Landscape Architects or ASLA's national office for information on landscape architects in your area.*

Ask to see examples of residential landscaping the individual or company has done. You may also want to visit a home where landscaping work is in progress to observe how the crew operates and how involved the designer is with the execution of his/her plans.

PLANNING THE LANDSCAPING

You will want to consider the following as you work with a specialist on the landscaping of your new home:

❏ How much maintenance will be required? Are you willing to spend the time and effort yourself, or would you prefer to retain a landscape maintenance service? (Your landscaper should give you care instructions for all plantings.) If you retain a service, how much will it cost, and what portion of the maintenance will you handle?

❏ What will the plantings look like? After you have reviewed the landscape plan, visit a nursery to look at examples of the plantings your designer has specified. Do you like their shape, smell, foliage, color? Are they evergreens, or are they deciduous trees that will drop their leaves each autumn? If so, does it matter to you?

❏ How big will the plantings grow, and how quickly? Will they look established and mature by your second year in the house? If not, what can be done to compensate for that bare look in the interim? Planting larger shrubs at the outset and annual flowering plants can fill in some of the bare area until other plantings mature.

---

* American Society of Landscape Architects, 1733 Connecticut Avenue, N.W., Washington, D.C. 20009, 202/466-7730.

- ☐ Can the plantings be pruned easily?
- ☐ What about a specimen tree or shrub? Decorative trees such as weeping willows, magnolias, a red maple, or a clump of birches can add color and focus to your yard.
- ☐ What about creating a vista with a special planting of shrubs around a sundial, planting a colorful and fragrant herb garden next to the patio, attracting birds with a nicely landscaped birdbath, or using a piece of sculpture as a focal point?
- ☐ Use plantings to define the head of a driveway, borders, or boundary lines between you and your neighbors; and shrubs to screen a fence, patio, deck, or pool.
- ☐ Lawns should integrate with the plantings. Perhaps you will want to keep grass to a minimum for aesthetic or maintenance reasons, and use walkways, small gardens, and shrub groupings instead. If you live in a rural area, you may want to experiment with a wildflower lawn. *
- ☐ If you are lucky enough to have your home ready for landscaping during a slow season for nurseries, you may be able to get a discount on plantings. Of course, you will want to plant during the temperate months according to the requirements of individual plant species. But buying plants in off months such as August or September gives the plants plenty of time to "take" before intemperate weather sets in.
- ☐ Did excavation turn up any boulders that can be worked effectively into the landscape plan? Could the topsoil be kept separate from the subsoil during excavation and saved for use in your gardens?

LANDSCAPE LIGHTING

If you work with a landscape architect, he/she should be an outdoor lighting expert. Or you can work with a lighting professional to "moonscape" your land. Moonscaping is just that: creating a beautiful night landscape, as though a full moon were hung over your yard. Custom designed night lighting avoids puddling and spotty effects. It highlights specimen plantings and other special features, illuminates paths with diffused light, and creates inviting views.

In addition to creating ambience, moonscaping can be planned to provide discreet security lighting without casting a glare on the neighbors or turning your driveway into daylight.

---

* Contact the American Horticultural Society, Box 0105, Mt. Vernon, VA 22121, 703/768-5700, for information on wildflowers and many other plants.

88

The cost of landscape lighting varies with the size and complexity of the project. A landscape architect may incorporate the cost into the overall landscaping fee, while a lighting professional will bill separately. Some lighting professionals will also maintain your lighting on a regular basis for an additional fee.

Chapter Ten

SETTLEMENT AND SETTLING IN

*Y*our house is finished. It has passed all final inspections and you are more than ready to move in. But first—a few details that must be attended to.

PRESETTLEMENT WALK-THROUGH

When you and your builder drew up your contract, you agreed on a procedure for final inspection. Now is the time for that inspection, or "walk-through," to take place. It should be scheduled far enough in advance of the settlement date to give the builder time to remedy any problems you may discover.

What are you looking for? Windows that don't open properly, missing items such as a towel bar or a light switch plate, a spot where the paint didn't cover, a sticking door, a chipped countertop, and so on.

The most efficient way to conduct a walk-through is to use a checklist such as the one shown in Figure 14. One signed copy of the completed checklist stays with you; the other copy goes to the builder. While builders prefer to remedy problems before you move in, some items may have to be corrected after settlement. For instance, if your walk-through takes place in the winter and landscaping adjustments are required, these will have to wait until spring. Touch-up on exterior trim may have to wait until spring, as paint will not adhere properly to cold or wet surfaces. You and the builder should agree on completion dates for such items. It is important to remember that the builder cannot be responsible for fixing any item not noted on your walk-through checklist.

14. Sample presettlement walk-through checklist.

Purchaser _____

Address _____

Inspection date _____ Occupancy date _____

       (builder)   is proud to welcome you to your new home. Attached is a checklist to help you inspect your new home. A separate sheet has been provided for each room and area of construction.

    Please go through your new home, room by room, and carefully check to see whether all items are in satisfactory condition. Initial the space provided after you have satisfied yourself that the item's condition meets with your approval. If any repairs or adjustments are needed, describe the problem in the "Needs Improvement" column, using the back of the page if necessary. Be sure to write in any additional items that are not listed, and describe their condition fully. If a listed item does not apply, put an "X" in the space for your initials.

    We will do our best to bring the items that require improvements up to satisfactory condition, consistent with the standards of construction in   (city, state)   , and with our Builder's Limited Warranty, before you occupy your home. A second inspection will be required at that time, and you will have the opportunity to check all the items to make sure they meet with your approval.

Item	Initials	Needs Improvement
Bathroom		
Vanity	_____	_____
Sink	_____	_____
Medicine cabinet	_____	_____
Bathtub	_____	_____
Shower	_____	_____
Shower curtain bar/door	_____	_____
Commode	_____	_____
Towel bars	_____	_____
Paper holder	_____	_____
Light fixture (not bulbs)	_____	_____
Radiator	_____	_____
Floors	_____	_____
Walls	_____	_____
Ceiling	_____	_____
Woodwork	_____	_____
Windows	_____	_____
Doors	_____	_____

Item	Initials	Needs Improvement
Kitchen		
Sink	_____	_____
Stove	_____	_____
Oven, range	_____	_____
Hood and exhaust fan	_____	_____
Appliances		
Microwave	_____	_____
Mixer	_____	_____
Dishwasher	_____	_____
Refrigerator	_____	_____
Freezer	_____	_____
Disposal	_____	_____
Cabinets	_____	_____
Drawers	_____	_____
Countertops	_____	_____
Floor	_____	_____
Walls	_____	_____
Ceiling	_____	_____
Light fixtures (not bulbs)	_____	_____
Windows	_____	_____
Doors	_____	_____
Radiator	_____	_____

Living Room, Dining Room, Den, Family Room, Bedrooms

Item	Initials	Needs Improvement
Doors	_____	_____
Windows	_____	_____
Shades	_____	_____
Closets	_____	_____
Light fixtures (not bulbs)	_____	_____
Floor	_____	_____
Walls	_____	_____
Ceiling	_____	_____
Vents	_____	_____
Radiator	_____	_____
Other items	_____	_____

Item	Initials	Needs Improvement
Hallway		
Closets	————————	————————————
Walls	————————	————————————
Floors	————————	————————————
Ceiling	————————	————————————
Light fixtures (not bulbs)	————————	————————————
Doors	————————	————————————
General Interior		
Interior doors	————————	————————————
Hardware	————————	————————————
Paneling	————————	————————————
Insulation	————————	————————————
Wallpaper, paint	————————	————————————
Fireplace(s)	————————	————————————
Cabinets, bookcases	————————	————————————
Woodwork	————————	————————————
Tile work	————————	————————————
Windows	————————	————————————
Doors	————————	————————————
Walls	————————	————————————
Glass	————————	————————————
Ceilings	————————	————————————
Other items	————————	————————————
General Exterior		
Paint	————————	————————————
Siding, brick	————————	————————————
Chimney	————————	————————————
Roof	————————	————————————
Doors	————————	————————————
Garage	————————	————————————
Walkways	————————	————————————
Balcony	————————	————————————
Light switches	————————	————————————

Item	Initials	Needs Improvement
Light fixtures (not bulbs)	_____	_____
Outlets	_____	_____
Shades	_____	_____
Vents	_____	_____
Radiators	_____	_____
Other items	_____	_____

I understand each of the items on the preceding pages and have discussed each item with a representative of _____ (builder) _____ . I have been instructed in the use and care of the items listed on the above pages. With the sole exception of the items listed under "Needs Improvement," I find my new home to be completed in a manner satisfactory and acceptable to me. I understand that, with the exception of those items that I have stated under "Needs Improvement," I am purchasing the house as is, and I understand that the builder makes no other guarantees or warranties other than those that are clearly stated in the Agreement and the other Contract Documents.

_____   _____
Buyer   Builder's firm

_____   _____
Buyer   Representative of firm

_____   _____
Date   Title of representative

                             _____
                             Date

If an architect designed your home, he/she should also make a final inspection and report any problems to you so that the builder can correct them prior to settlement.

BUILDER WARRANTY

Whether the builder is a member of an insured warranty plan or offers an uninsured warranty, coverage goes into effect at closing. The builder should review the warranty with you prior to settlement. Read all warranty documents carefully and be sure you understand the terms of the protection that is offered.

SETTLEMENT

Settlement (or closing) is the process whereby ownership of a property is passed from seller to purchaser. In the case of a custom home, your lender may be the "seller." More commonly, the builder is the seller because he/she has put together the land, construction, and permanent financing package. In this instance, final payment to the builder is made at settlement and the keys to the house are transferred from builder to owner.

The following individuals are generally present at closings: you, your attorney, the builder, a representative of the lending institution that is financing your mortgage, and a title company representative. No matter how your financing is structured, settlements involve costs and the presentation of certain documents. These are described below.

Settlement costs

❏ *Points.* If you are paying points, they may add substantially to your settlement costs. As Chapter 4 described, points are levied by the lending institution to increase its yield on a loan. Each point equals one percent of the loan's total. Points are payable in full at closing.

❏ *Appraisal.* An appraisal is an estimate of your home's fair market value. Lenders require an appraisal to determine how much they could recover by selling your house if you default on your loan; they will probably charge you for the service. An appraiser inspects the house and the neighborhood, basing his/her estimate on the selling price of comparable houses in the area and other factors. Some banks have their own appraisers; others use outside firms. Fees vary.

❏ *Credit report.* The lender may charge a fee for investigating your credit history.

❏ *Mortgage origination fee.* The lender charges a fee for processing the loan, usually calculated as a percentage of the loan amount.

❏ *Property survey fee.* Your lender will require a survey as part of the mortgage loan package. The fee is your responsibility.

❏ *Title search and insurance.* You will have had a title search prior to buying the land, but the lender will require a title insurance commitment at closing to determine that no mechanics liens, mortgages, or other claims have been placed since the initial title search. If the same company performs both the original title search and this update, the charge should be lower. You will usually be required to purchase lenders' title insurance to guard against a faulty title search and other unexpected problems. Owners' title insurance is also available to protect you against legal expenses or the value of property lost due to title problems. Title insurance is discussed further in Chapter 6.

❏ *Taxes, fees, and other charges.* Some jurisdictions levy taxes on the transfer of property or on real estate loans, which will be due at closing. The lender will advise you of any such requirements. If your builder paid property taxes on the house or filled the fuel tank, you must pay for a proportional share of the taxes, any fuel remaining in the tank, and other prepaid costs.

❏ *Insurance.* In most cases, homeowners' insurance with a fully paid first year's premium is required at settlement.

❏ *Lawyer's fee.* It is recommended that you retain a real estate attorney to review your contract and settlement documents, and to represent you at closing. The legal fees for a residential real estate closing are relatively inexpensive ($200-$400) and are well worth the investment. If your lender uses an attorney to draw up any of the settlement papers, you may be charged an additional fee—usually a flat amount, which will be part of the closing costs.

Under the Real Estate Settlement Procedures Act (RESPA), your lender is required to give you a copy of the *Settlement Costs* booklet issued by the U.S. Department of Housing and Urban Development (HUD).* This booklet will help you figure out settlement costs prior to the closing. RESPA also requires your lender to give you a cost rundown.

Settlement documents

There are certain documents you are required to supply at closing, and others that you should receive at that time.

Documents you supply

❏ Certificate of occupancy from the building inspector indicating that the building is habitable (required in some jurisdictions).

❏ Inspection certificates from the building and health departments for code compliance.

❏ Boundary survey.

❏ All required insurance policies.

Documents you receive

❏ Copy of the sales contract, if you are purchasing the house from the builder.

❏ All applicable tax payment receipts from the builder.

❏ Mechanics lien release from the builder, subcontractors, and suppliers releasing you, the owner, from lien liability.

❏ The note and deed to your property (which will probably be mailed to you after being recorded in your local registry of deeds office).

---

* *Settlement Costs [HUD-398-H(3)].* U.S. Department of Housing and Urban Development, Washington, D.C., 1987. This booklet is available from the Superintendent of Documents, U.S. Government Printing Office, Washington, D.C. 20402.

❑ Home maintenance and care instructions from your builder, as well as manufacturers' warranties and instruction booklets for equipment in the house. Warranties on appliances and other items go into effect at date of installation and are therefore installed as close to the settlement date as possible. Keep copies of each warranty in a safe place. It is your responsibility to register your name and the model number of the appliance with the manufacturer. You will have to negotiate any claims on appliances unless the builder specifically agrees to do this.

❑ Schedule of mortgage payments (amortization schedule) which will show the amount of interest and principal to be paid each month. If you are required to pay taxes and insurance as part of the monthly payment, these will also be indicated. With an adjustable rate mortgage (ARM), the amortization schedule will change periodically with every mortgage interest rate change. Your lender will supply you with revised schedules.

SERVICE

As with walk-throughs and warranties, the builder's service policy should have been clearly stated at the time the contract was drawn up. Many builders give owners a chance to get acquainted with their new homes, then schedule routine callback visits 30 to 60 days after closing and again after 11 months. The builder should give you a telephone number to call for service requests. You should understand, however, that the builder will not send out a service person every time a problem is discovered. Compile a list of items that can be taken care of during a regularly scheduled service visit. Of course, the builder will handle emergencies as promptly as possible. Remember that the perfect house will probably never be built—even if it is your dream house.

What is your recourse

Suppose your builder cannot or will not take care of problems that arise with your new home. What should your course of action be? First, carefully read through your warranty policy. Problems covered under the warranty should be easily correctable. If a dispute over warranty coverage arises between you and the builder, an impartial third party will be brought in by the warranty company to mediate the dispute. If the problem develops after the warranty has expired, however, the builder is not required by law to correct the problem. (Also remember that some items, such as appliances, may be covered by a manufacturer's warranty and are not the builder's responsibility to maintain.)

Next, identify the exact nature of the problem in a letter to your builder. If you have an insured warranty that covers the problem, send a copy of the letter and attachments to the warranty company as well. Use the following guidelines in writing your letter:

❏ Include your name, address, and home and work telephone numbers.
❏ Type the letter, if possible. If not, write legibly.
❏ Make the letter brief and to the point, but include all relevant details.
❏ State exactly what you want done and how soon you expect the problem to be resolved. Be reasonable.
❏ Include copies of all documents related to the problem. Do not send originals.
❏ Keep a copy of the letter and attachments for your records.

Do not send letters to lawyers, government agencies, the local builders association, the Better Business Bureau, or your local consumer action reporter before you have given the builder and the warranty company a reasonable chance to resolve the problem. Contact outsiders only after you have reached an impasse with the builder. The arbitration process, which is described in Chapter 6, is preferable to legal proceedings. Going to court tends to be expensive and time-consuming.

Of course, chances are the builder will take care of all legitimate problems and see to it that your home is everything you dreamed it would be. Most builders count on referrals from satisfied customers to keep their businesses growing.

MOVING IN

No matter how well organized you are, moving is an ordeal involving endless details and decisions. Use the following checklist to help you prepare for moving day.

2-3 months before move

❏ Start contacting moving companies for bids. If you are moving during the summer months, allow extra time to line up movers as this is their busiest time. Once you have chosen a moving company and confirmed the moving date with them, decide whether you or professional packers will pack your belongings.

6 weeks before move

❏ Begin notifying magazines, credit card companies, and others of your address change. Keep a list of everyone you notify.

❑ Start a file or notebook of important papers and "things to do." Organize chronologically or by subject.

5 weeks before move

❑ Begin sorting through your belongings, deciding what to keep, what to give to a rummage sale, what to sell at a garage sale.

❑ Start packing little-used and off-season items. Designate an area in the house for storing packed boxes. Label each box with contents and destination (say "Bedroom 1" rather than "Jimmy's Room").

❑ Do an inventory of your household goods, listing value or original purchase price, date and place of purchase (you may already have such an inventory for insurance purposes). You may want to photograph large or especially valuable items. Should anything be damaged during the move, this will be important information to have.

❑ Begin planning the placement of furniture in your new home. If you are working with an interior designer, you will receive a floor plan showing the furniture arrangement for each room. This will make the move-in process go faster and more smoothly for both the movers and you.

4 weeks before move

❑ Fill out a change-of-address card at the post office, to take effect on move-in day. (This can be done earlier if you are positive of the date you will move into the new house.)

❑ If you will be moving some distance away, now is the time to talk with your bank, insurance agent, and attorney about closing accounts, changing policies, and any legal or contractual matters that need to be resolved before you move.

3 weeks before move

❑ If you are moving to a new area, take care of medical arrangements. Fill prescriptions and get a copy from your doctor for your new pharmacy; ask your doctors and dentists to recommend colleagues in your new locale; get copies of medical and dental records.

❑ Get copies of school transcripts and other school records.

2 weeks before move

❑ Arrange for servicing and special packing of appliances, piano, computer, and other large items to be moved. You will want to have these items checked once they are installed in your new home.

- [ ] If you are doing your own packing, the time has come to start packing more frequently used belongings. Again, label each box with contents and destination.
- [ ] Reconfirm move date with moving company.

One week before move

- [ ] Start to empty out the kitchen cabinets and refrigerator by planning easy meals that use up leftovers and require few pots and pans.
- [ ] Arrange to have gas, electric, telephone, water turned on in the new house by moving day.
- [ ] Return library books and other borrowed items; collect items others have borrowed from you.
- [ ] Arrange to have newspaper delivery cancelled at your old address (and started at your new address, if you wish) as of moving day.
- [ ] If you will be moving nearby, arrange for a pet sitter and child care on moving day.
- [ ] If you will be driving a distance to your new home, get your car and routing checked.
- [ ] Empty fuel from lawnmower, gas grill, snowblower, and other equipment in preparation for the move.

2 days before move

- [ ] Reconfirm move date with moving company.

One day before move

- [ ] Pack all last-minute kitchen and bath items.
- [ ] If you are moving a distance away, pack clothing and household essentials (trash bags, soap, paper towels, magic marker, string, etc.) to use until movers arrive at the new house.
- [ ] Empty out, defrost, and clean refrigerator.
- [ ] Load up the car with plants and any other special items you will be moving yourself.

Moving day

- [ ] Sweep and vacuum.
- [ ] Strip beds.
- [ ] Check all cabinets, closets, drawers, the dishwasher, washer and dryer one last time.

- Leave new owner the house keys and your new address so that any mail can be forwarded.
- Also leave appliance maintenance booklets and warranties for the new owner.
- As a courtesy, you may wish to leave a list of neighbors' names, and telephone numbers for police, fire department, etc.
- Be on hand to answer movers' questions, and give them a telephone number where you can be reached until they arrive at your new house.

A word about movers: They are professionals, so you should not hover over them as they do their job. However, you will want to supervise the moving of any precious or fragile items; it seems as though the only things that ever get broken are the ones that cannot be replaced. Let the movers know beforehand about any special packing requirements you may have: paintings, heavy mirrors, tall clocks, chandeliers and the like that require special handling.

If you have fine antiques, retain a mover who specializes in moving valuable pieces. Any pieces that require repair or reupholstering can be sent out during construction and delivered directly to the new home.

No matter how well boxes are labeled, the movers will need instructions on where to put them. Keep the floor plan at hand to direct the placement of boxes, rugs, and furniture.

Supervise the unpacking of special items, but plan to put them in place yourself unless you've made special arrangements with the movers. Normally, this is not part of their work.

Remember that the builder is not responsible for repairing damage caused by the movers. Scratches in your new hardwood floor and nicks in that fresh paint job will have to be fixed by you.

Most movers will require payment on the spot by bank or certified check. Have a check ready in the proper amount to guarantee that the furniture will be offloaded from the truck.

Welcome to your dream home.
May it be filled with joy.

APPENDIX

State Builders Associations Affiliated With the National Association of Home Builders

The National Association of Home Builders (NAHB) is a federation of more than 800 state and local builder associations representing 150,000 builders and related industry professionals throughout the country.

The following is a list of state builders associations, listed in alphabetical order by state. Your state association can refer you to the local builders association for your area. You can also contact the National Association of Home Builders for more information on local builders associations. (National Association of Home Builders, 15th and M Streets, N.W., Washington, D.C. 20005, 202/822-0200.)

Home Builders Association of Alabama
Montgomery, AL   205/834-3006

Alaska Home Builders Association
Anchorage, AK   907/522-3605

Home Builders Association of Central Arizona*
Phoenix, AZ   602/274-6545

Flagstaff League for Advancing Good Growth*
Flagstaff, AZ   602/526-4464

Southern Arizona Home Builders Association*
Tucson, AZ   602/795-5114

Arkansas Home Builders Association
Little Rock, AR   501/663-1428

California Building Industry Association
Sacramento, CA   916/443-7933

Colorado Association of Home Builders
Denver, CO   303/753-0601

Home Builders Association of Connecticut
Hartford, CT   203/247-4416

Home Builders Association of Delaware
Wilmington, DE   302/994-2597

District of Columbia Building Industry Association
Washington, D.C.   202/966-8665

---

*Arizona has no state builders association.

Florida Home Builders Association
Tallahassee, FL   904/224-4316

Home Builders Association of Georgia
Atlanta, GA   404/763-2453

Building Industry Association of Hawaii
Honolulu, HI   808/847-4666

Idaho Building Contractors Association
Boise, ID   208/384-5503

Home Builders Association of Illinois
Springfield, IL   217/753-3963

Indiana Builders Association
Indianapolis, IN   317/236-6334

Home Builders Association of Iowa
Des Moines, IA   515/278-0255

Home Builders Association of Kansas
Topeka, KS   913/233-9853

Home Builders Association of Kentucky
Frankfort, KY   502/875-5478

Home Builders of Louisiana
Baton Rouge, LA   504/387-2714

Home Builders Association of Maine
South China, ME   207/445-4590

State of Maryland Institute of Home
Builders
Annapolis, MD   301/261-2997

Massachusetts Home Builders Association
Boston, MA   617/720-2340

Michigan Association of Home Builders
Lansing, MI   517/484-5933

Builders Association of Minnesota
Saint Paul, MN   612/646-7959

Home Builders Association of Mississippi
Jackson, MS   601/969-3446

Home Builders Association of Greater
Kansas City*
Kansas City, MO   816/942-8800

Home Builders Association of Greater St.
Louis*
St. Louis, MO   314/994-7700

Montana Building Industry Association
Helena, MT   406/475-3722

Nebraska State Home Builders
Association
Omaha, NE   402/333-7731

Nevada Home Builders Association
Las Vegas, NV   702/794-0117

Home Builders Association of New
Hampshire
Concord, NH   603/228-0351

New Jersey Builders Association
Plainsboro, NJ   609/275-8888

New Mexico Home Builders Association
Albuquerque, NM   505/344-7072

New York State Builders Association
Albany, NY   518/465-2492

North Carolina Home Builders Association
Raleigh, NC   919/833-4613

North Dakota Association of Builders
Bismarck, ND   701/222-2401

Ohio Home Builders Association
Columbus, OH   614/228-6647

Oklahoma State Home Builders
Association
Oklahoma City, OK   405/843-5579

Oregon State Home Builders Association
Salem, OR   503/378-9066

Pennsylvania Builders Association
Harrisburg, PA   717/234-6209

Home Builders Association of Puerto Rico
Hato Rey, PR   809/751-1471

Rhode Island Builders Association
Providence, RI   401/521-0347

---

*Missouri has no state builders association. Either of these associations can refer you to the local builders association in your area.

Home Builders Association of South Carolina
Columbia, SC   803/771-7408

Home Builders Association of South Dakota
Sioux Falls, SD   605/361-8322

Home Builders Association of Tennessee
Nashville, TN   615/726-1700

Texas Association of Builders
Austin, TX   512/476-6346

Home Builders Association of Utah State
Salt Lake City, UT   801/268-8750

Home Builders Association of Vermont
Williston, VT   802/879-7766

Home Builders Association of Virginia
Richmond, VA   804/643-2797

Building Industry Association of Washington
Olympia, WA   206/352-7800

Home Builders Association of West Virginia
Charleston, WV   304/342-5176

Wisconsin Builders Association
Madison, WI   608/249-9912

Wyoming Home Builders Association
Cheyenne, WY   307/632-5557

GLOSSARY

Adjustable rate mortgage (ARM) A form of loan where the interest rate is adjusted according to rate fluctuations in the overall financial market. The consumer generally receives a lower initial interest rate from the lender, but assumes part of the risk of an increase in interest rates. The rate adjustments are limited by a **cap** set by the lender.

AIA See **American Institute of Architects**.

Allowance On a builder's estimate, an item for which "x" dollars are allowed. Should you select another make or model costing more or less, the builder will adjust the estimate accordingly.

ALTA See **American Land Title Association**.

American Institute of Architects (AIA) A professional membership society for licensed architects (American Institute of Architects, 1735 New York Avenue, N.W., Washington, D.C. 20006).

American Land Title Association (ALTA) A membership organization of title insurers, agents, attorneys, and others specializing in real estate law (American Land Title Association, 1828 L Street, N.W., Washington, D.C. 20036).

American Society of Interior Designers (ASID) A professional membership society for interior designers (American Society of Interior Designers, 1430 Broadway, New York, New York 10018).

Amortization A payment plan in which a borrower gradually reduces a debt through regularly scheduled payments of principal and interest.

Annual percentage rate (APR) The annual cost of credit over the life of a **mortgage** loan. It includes **interest**, service charges, **points**, loan fees, **mortgage insurance**, and other charges, and is therefore higher than the base **interest** rate. It allows for "comparison shopping" among **mortgages** with different terms and features.

Appraisal An estimate of your home's fair market value, based on the selling price of comparable houses in the area and other factors.

Apprenticeship A training period during which a person works under an experienced practitioner to gain skills in a particular discipline.

APR See **annual percentage rate.**

ARM See **adjustable rate mortgage.**

ASID See **American Society of Interior Designers.**

Asphalt A dark, tarlike material commonly used in the building industry for such uses as roofing, waterproofing and dampproofing, exterior wall covering, and pavement. See also **bituminous compound.**

Backfill Earth or other material used to fill in around foundation walls, usually built up to drain water away from the foundation.

Backsplash The continuation of a countertop surface part way up a wall to protect the wall against water damage.

Balloon mortgage A type of **mortgage** loan that offers equal monthly payments and a single large final payment to pay off the loan within a set number of years.

Baseboard A decorative band of finish board used to cover the joint between wall and floor.

Bid An architect, builder, interior designer, **subcontractor,** or other supplier's proposal to work on a project; generally includes price estimate and description of work to be performed.

Binder A sum of money paid to an architect, builder, interior designer, **subcontractor,** or other supplier to reserve their services until a formal contract or agreement has been signed.

Bituminous compound A dark, tarlike material commonly used throughout the building industry for such uses as roofing, waterproofing and dampproofing, exterior wall covering, and pavement. See also **asphalt.**

Blueprints Complete construction plans, drawn to scale, used by builders and **subcontractors** to build a structure. They usually include **site plan, foundation** plan and section, **floor plans, elevations,** building and wall **sections, mechanical systems,** and special construction details. The term blueprint derives from the chemically treated blue paper on which drawings are printed in white; the term commonly refers to any set of working drawings, whether printed on blue, white, or other paper.

Builders association Any of more than 800 state and local membership affiliates of the National Association of Home Builders (NAHB), whose members are builders and related industry professionals. A complete list of state builders associations appears in the Appendix of this book.

Builders' risk insurance Insurance on construction to protect against vandalism, theft, structural collapse, and other building site hazards.

Building code Minimum legal requirements for all aspects of construction, established and enforced by local governments to protect public health and safety.

CAD See **computer assisted design.**

Callback   Request by a homeowner for a builder to handle a service request.

Cap   In finance, the maximum percentage that the interest rate on an **adjustable rate mortgage** can be changed, both in a single adjustment period and over the life of the loan.

Cement   A limestone- and clay-based substance applied wet to bind together materials such as sand, gravel, brick, and stone when it hardens. It is an ingredient in **mortar** and **concrete**. The term cement is often incorrectly used to mean **concrete**.

Center beam   A wood or steel member that runs the length of the first floor of a house, bearing on the foundation wall at each end of the house and supported along its length by columns or piers. The center beam supports the house structure above it.

Certificate of occupancy   A legal document issued by a building inspector, stating that a house has passed all **inspections** and is ready for occupancy.

Chair rail   A band of **molding** applied at chair back height along a wall to protect the wall finish from chairs being pushed against it.

Change order   Home buyer's written authorization to add, delete, or change an item specified in a contract.

Closing   The process by which a house purchase transaction is finalized. It involves presentation, review, and sign-off on a number of legal documents and the payment of certain closing costs. Also called settlement.

Code   See **building code**.

Collateral   Personal property used to secure or guarantee a loan.

Commercial bank   A financial institution that formerly specialized in short-term loans, but which is becoming more active in the consumer loan market, including home mortgages.

Commission   In real estate, the percentage of the total selling price of a property paid to the salesperson who handles the sale.

Comprehensive plan See **master plan**.

Computer assisted design (CAD) Use of a computer and software to prepare architectural and construction plans and drawings.

Concrete   Mixture of cement, sand, gravel, and water that hardens into a rocklike mass. The term concrete is often incorrectly used to mean **cement**.

Contingency clause   Terms added to a **contract** that must be met in order for the contract to become binding.

Contract   A legal agreement between two parties specifying terms and conditions under which a certain type of work is to be performed.

Contractor   The individual or company who is responsible for overall construction of a project. Negotiates, schedules, and manages work of **subcontractors**. Also called general contractor.

Cornice On the exterior of a house, construction under the **eaves** or where the roof and walls meet. Also, a projecting decorative **molding** that crowns or finishes the top of a wall or building.

Cost-plus A type of contract or agreement where a builder or architect bills the customer for materials and labor at cost, and receives a set fee or percentage of cost.

Course A horizontal row of masonry, such as brick or **concrete** block.

Covenant An agreement between the seller and purchaser of a piece of property, restricting the use of that property. Covenants may also be established by neighborhood associations, developers, or local governments to protect or preserve a desirable element of the community. Also called deed restriction.

Crawl space In houses without basements, the space between the ground surface and the first floor, made big enough to "crawl around in" for utility installation and repairs.

Credit union A cooperative savings association that makes low-interest loans to its members.

Cross section An architectural or construction drawing that has been cut vertically to reveal the structure behind the surface. Also called section.

Crown molding A decorative band of finish board used to cover the joint between wall and ceiling.

Cultured marble A manufactured marblelike material commonly used for countertops and for lavatory surfaces. It is a cast polyester resin mixed with crushed marble, then molded, cut, and polished. It is water resistant, and is lighter weight and less expensive than **quarry marble**.

Deciduous In landscaping, a tree or shrub that loses its foliage seasonally.

Deductible A clause in an insurance policy that stipulates what portion of a claim the policy-holder must pay.

Deed A legal document representing property ownership.

Deed restriction See **covenant**.

Default Failure to make payment on a loan when payment is due.

Density The allowable number of persons, families, or structures per unit of land as established by **zoning ordinance**.

Dormer A projection built out from a sloping roof as a room extension or for a window.

Double-glazed window A window containing two sheets of glass with an air space sealed in for insulation purposes.

Downdraft stovetop A cooking unit with ventilation built into the stove surface.

Downpayment An initial payment on the purchase of a house, usually paid at **closing**, with the balance of the purchase price paid in monthly installments against a **mortgage** loan.

Draw   In financing, a scheduled construction loan payment.

Drywall   General term for a type of interior wall construction using "dry" **gypsum wallboard** panels instead of plaster; also called sheetrock, wallboard.

Duct, ductwork   Round or rectangular sheet metal or vinyl pipe used to transfer heated and cooled air from heat and air conditioning sources to the various rooms in a building.

Easement   A right given by the owner of a piece of land to an individual, government, or other entity for a specific limited use of that land.

Eave   That part of a roof that extends beyond the walls of a building.

Elevation   Vertical scale drawing of an interior or exterior building surface.

Encroachment   An improvement to property that illegally extends beyond that property's lot line.

Equity   The value of property beyond what is owed in mortgage payments to a lender.

Excavation   Removal of earth or rock to create a hole, as for the basement of a house.

Façade   The exterior surface or "face" of a building, usually the front.

Fannie Mae   See **Federal National Mortgage Association**.

Federal Home Loan Mortgage Corporation   A federally chartered institution that participates in the secondary mortgage market by purchasing conventional loans and **FHA-** and **VA**-insured mortgages. Also called FHLMC or "Freddie Mac."

Federal Housing Administration (FHA)   A federal agency that insures home **mortgages**, allowing persons who qualify to purchase homes at a lower **downpayment** than is possible with a conventional loan. The federal government limits the amount that can be borrowed with an FHA-insured loan.

Federal Insurance Administration (FIA)   An office of the Federal Emergency Management Agency that prepares and distributes flood hazard maps for insurance purposes (Federal Insurance Administration, 500 C Street, S.W., Washington, D.C. 20472).

Federal National Mortgage Association   A federally regulated institution that buys and sells residential mortgages in the secondary mortgage market. Also called FNMA or "Fannie Mae."

FHA   See **Federal Housing Administration**.

FIA   See **Federal Insurance Administration**.

Fiberglass   A nonflammable material made of spun glass fibers. It is used in thick woollike blankets as building insulation; woven into fabrics; and used to reinforce plastic resins in durable, molded, solid forms for a variety of uses.

Fixed rate mortgage   A **mortgage** loan with non-adjustable monthly payments based on **interest** rates at the time the loan is initiated.

Flakeboard A **plywood** substitute manufactured from wood flakes and a resin binder pressed into boards.

Flashing Sheet metal or plastic used to cover joints and openings in exterior surfaces to protect against water leakage.

Floor plan An architectural drawing of each floor of a structure, drawn in scale to show location and dimensions of all rooms, windows, doors, walls, stairs, and fixtures.

Footing Widened support, usually concrete, at the base of **foundation** walls, columns, piers, and chimneys. Designed to distribute the weight of these elements over a larger area and prevent uneven settling.

Foreclosure In the event of a borrower's failure to make a scheduled payment on a **mortgage**, a lender's action that deprives the borrower of the right to redeem the mortgage; ownership of the property then passes to the lender.

Formwork Support system for freshly placed concrete.

Foundation The entire substructure, usually below ground, on which a building rests.

Foyer Interior entrance hallway or area of a building.

Framing The process of constructing the internal skeleton of a structure, usually of wood or steel **studs**, beams, and **joists**.

Freddie Mac See **Federal Home Loan Mortgage Corporation.**

French door A doorway consisting of a pair of center-opening doors with glass panels extending the full length of the doors.

General contractor See **contractor.**

Ginnie Mae See **Government National Mortgage Corporation.**

Government National Mortgage Corporation A federal government agency that guarantees payment to investors on **FHA** and **VA** loans sold to the **secondary market.** Also called GNMA or "Ginnie Mae."

Grading The preparation of a site by digging, filling in, or both, to accommodate construction of a building. Also, filling in with earth or other material around a completed building, at a slope to direct rain water runoff away from the building.

Gutter Metal, plastic, or wood channel at the **eaves** of a building, sloped slightly to carry off rain water and snow melt.

Gypsum board Panels used in **drywall** construction, consisting of the mineral gypsum pressed between two layers of heavy paper. Also called sheetrock, wallboard.

Hearing A community planning commission meeting at which requests for **variances** and other deviations from **zoning ordinances** are considered.

Homeowners' insurance A type of insurance policy carried by homeowners, covering damage, fire, theft, and injury.

HVAC Common building industry abbreviation for heating, ventilation, and air conditioning systems.

Indexing In finance, the periodic adjustment of interest rates according to standards such as U.S. Treasury bills.

Inspection Examination of work completed on a structure to determine compliance with **building code** and other code requirements.

Insulation Any material used in building construction to resist heat loss, protect against sound transmission or fire, or to cover electrical conductors.

Interest In finance, the rate paid to a lender for the use of money borrowed; usually expressed as a percentage.

Joint compound A pastelike material used to cover tape at drywall seams for a smoother finish. Also called spackle.

Joists A series of horizontal parallel beams that support floors and ceilings.

Liability insurance Financially protects the insured individual in the event of injury or property damage against another.

Lien A claim on property as security for payment of a debt.

Load bearing Providing support for a building's weight.

Masonry General construction term for materials set in **mortar**, including stone, brick, concrete, tile, and glass block.

Master plan A document prepared by a community's planning department that identifies future growth areas in that community and general development goals in such areas as housing, transportation, schools, health care, and public **utilities**. Also called comprehensive plan.

Mechanics lien A type of **lien**, or claim, on a property for payment of a debt related to that property, by persons supplying labor or materials for the construction of that property.

Membrane waterproofing Method of protecting a structure against moisture transfer using **polyethylene**, felt layers, or other water-resistant material.

Mineral wool A loose, fibrous **insulation** material made from rock and molten slag.

Molding Wood, metal, or plaster strips used for decorative finish around windows and doors, at the top and base of walls, and along **cornices**.

Mortar A thick, pastelike material that hardens to bond **masonry** units together. Usually made of a mixture of **cement**, lime, sand, and water.

Mortgage A type of loan made for the acquisition of property, in which the house and land are used as security for the loan. The borrower agrees to pay a lender a specific **interest** rate on money borrowed for a specific period of time; the money is repaid to the lender in installments. The borrower agrees to repay **principal** and **interest**, to keep the property insured and in good condition, and to pay all taxes. Also called permanent financing.

Mortgage insurance A form of insurance carried by a borrower to protect the lender financially in case of default on the loan.

NAHB See **National Association of Home Builders.**

National Association of Home Builders (NAHB) A professional membership organization for builders and remodelers involved in residential and light commercial construction, and for professionals in related industries (National Association of Home Builders, 15th and M Streets, N.W., Washington, D.C. 20005).

National Kitchen and Bath Association (NKBA) A professional society that certifies interior designers who specialize in kitchen and bath design (National Kitchen and Bath Association, 124 Main Street, Hackettstown, NJ 07840).

NKBA See **National Kitchen and Bath Association.**

Ordinance A regulation usually established and monitored by a local government.

Oriented strand board A **plywood** substitute composed of layers bonded together with resin. Each layer consists of compressed strands of wood fiber oriented in a single direction; layers alternate direction of strand orientation.

Pass-through A countertop or other surface that is open to two rooms, such as the kitchen and dining room, allowing items to be conveniently passed from one room to the other.

Percolation test A soil test used to determine the rate at which water will be absorbed into the ground. Results are used to establish best locations for **septic** fields on a piece of property and to determine their size. Also called "perc test."

Permanent financing See **mortgage.**

Permit A document issued by a local government agency allowing construction work to be performed in conformance with local **codes.** Work may not commence until permits have been obtained, and each permit-issuing agency must **inspect** the work at certain specified points during construction.

Plywood A type of building material made by gluing three or more thin layers (or "plies") of wood together in panels. Plies are laid so that the wood grain alternates direction with each layer; this increases the plywood's overall strength and counteracts warping in each ply.

Pocket door A door that slides into a "pocket" in the wall when opened.

Points In finance, one point equals one percent of the total amount of a loan. Points are a one-time charge assessed by a lender at **closing** to increase the yield on the loan.

Polyethylene A durable, pliable, waterproof plastic film available in 4- and 6-mil thicknesses, used in construction as a vapor barrier.

Presettlement walk-through Final inspection of a house prior to **closing,** conducted by builder and buyer.

Principal In finance, the total amount borrowed in a loan.

Purchase money mortgage A type of financing arranged between the buyer and seller of property, where the buyer gives the seller an agreed upon **downpayment** and then makes payments to the seller until the purchase price is paid off.

Quarry marble Marble that has been extracted from a naturally-occurring land source and has been cut and polished for use in construction.

R-value A term which, when used with a number, indicates the level of resistance to heat flow in a building material. The higher a material's R-value, the more effective **insulation** it provides.

Radon A colorless, odorless, tasteless radioactive gas that occurs naturally in soil, underground water, and outdoor air. It is found in varying levels throughout the United States.

Rendering An interpretive sketch or drawing depicting a designer's conception of a finished project.

Rough-in The stage of construction that follows **framing**, when installation of all systems that will be concealed behind the walls—plumbing, electrical wiring, and heating/ventilation/air conditioning—occurs.

Secondary market Savings institutions, mortgage bankers, federal agencies, and other institutions that buy and sell mortgages from primary lenders. The secondary market helps to keep sufficient funds in circulation to meet consumer mortgage demands.

Section See **cross section**.

Septic system A sewage disposal system for individual homes. A holding tank for raw sewage is installed in the ground, where sewage is broken down and liquefied by bacterial action. Small amounts of solid matter do not break down, but settle to the bottom of the tank. (The tank must be cleaned out every few years.) Liquid waste is discharged to a distribution or absorption field where it slowly passes into the soil and is purified.

Setback The minimum allowable distance between a structure and its lot lines.

Settlement See **closing**.

Sewer A system of pipes for carrying away storm runoff, waste water, or sewage to a municipal processing plant.

Shake Hand-split wood shingle.

Sheathing Sheets of **plywood, flakeboard, oriented strand board,** or **insulation** board, usually 4' x 8', used to cover the exterior of a building's framing.

Sheetrock See **drywall**.

Shingles Roof or wall covering of **asphalt**, wood, tile, slate, or other material cut into standard lengths, widths, and thicknesses.

Siding The exterior finish of a house applied over the **sheathing**; generally wood, plastic coated wood, vinyl, aluminum, or steel.

Signature loan A type of bank loan based on the borrower's ability to pay and guaranteed by the borrower's signature. The bank decides how much it can lend based on the borrower's income, credit history, and bank record. If the borrower fails to pay back the loan, the bank may use the borrower's personal assets for the money owed.

Sill A support member laid flat on the top of the **foundation** wall, used as the base for floor **framing**; also called the sill plate. Also, the member forming the lower side of an opening, such as a windowsill or doorsill.

Site plan View of a building site that shows existing features and the location of the proposed structure on the lot. Basic lot and building dimensions are indicated.

Slab A flat layer of plain or reinforced concrete.

Soil bore Hole drilled in soil to determine types of soil or rock support present in order to evaluate stability and load-bearing capacity of a piece of land.

Spackle A patching compound used to fill plaster or **drywall** cracks or nail holes; also called joint compound.

Specifications A contractual document describing in detail the work to be performed; method of construction; standards of workmanship; quality, type, and manufacturer of materials and equipment for a particular project.

Stakeout Measuring of house dimensions on a lot in accordance with the house plans and using stakes to indicate each corner.

Stucco A plaster **cement** used as a covering for exterior wall surfaces.

Stud Upright wood or metal members used as supporting elements in walls and partitions.

Subcontractor Someone who contracts with a **general contractor** to perform work on a specific part of a construction job, such as **excavation**, plumbing, electrical work, or landscaping. Also called "sub."

Subflooring Rough boards, **plywood, flakeboard,** or **oriented strand board** laid on top of the floor **joists**, to which the finish floor is fastened.

Subordinate In finance, to give up first claim to property in case of default.

Sump A hole designed to collect seepage water in a basement. When the collected water reaches a certain level, it is automatically pumped out with a "sump pump."

Survey A plan or map of a land area or building lot, showing dimensions, location of property corners, topography, and locations of improvements, **easements,** and **encroachments**.

Term insurance Insurance that covers the full amount of a **mortgage** and decreases in value and premiums as the **principal** is paid off.

Termite A wood-devouring insect that can demolish the woodwork of a structure.

Thrift institution A financial institution that makes **mortgage** and other longer-term loans. Also called savings and loan, mutual savings bank.

Title Legal document indicating right of ownership of property.

Title insurance Insurance against any **title** defects that may exist prior to the time the title is passed from one owner to the next, and which may come to light in a future transaction.

Toe molding A thin **molding** strip, such as a quarter round, nailed to the **baseboard** where it meets the floor. Also called shoe mold.

Trench A narrow **excavation** in the earth for the installation of pipes, drains, and cables.

Trim Interior finish materials, including door locks, knobs, hinges, and other metal hardware; moldings around windows and doors; and other decorative work.

Triple-glazed window A window containing three sheets of glass with air spaces sealed in for insulation purposes.

Turnkey Arrangement whereby a builder completes a house to the point at which it is ready for the customer to move in.

Underlayment Moisture resistant material, such as an **asphalt**-based paper, applied over roof and wall **sheathing** and under roof and exterior finish to prevent water from entering the structure; also used between subflooring and finish floors.

Utilities Public services available to all citizens of a community, such as water, electricity, gas, and sewage disposal.

Vanity A dressing table; also, a wash basin with an enclosed cabinet below.

Vapor barrier Treated paper or plastic film that retards the flow of air and moisture.

Variance Permission for a land use or lot **setback** that does not conform to a strict interpretation of local **zoning** regulations.

Veterans Administration (VA) A federal agency that administers veterans affairs, including insured loans to American veterans and their spouses.

Wallboard See **gypsum board, drywall**.

Warranty Insured or uninsured protection provided by a builder, remodeler, or other supplier, covering workmanship and materials for a specified length of time.

Workers Compensation Insurance, paid by the employer, granted to workers and their dependents against injury and death in the course of employment.

Zoning Division of a county or municipality into land use categories, and the establishment of regulations governing the use, placement, spacing, and size of land parcels and buildings in each category.

RESOURCES

Publications

The Apple Corps Guide to the Well-Built House, by Jim Locke. Boston: Houghton Mifflin Company, 1988.

Bathroom Design, by Jane Moss Snow. Washington, D.C.: National Association of Home Builders, 1987.*

Builder's Guide to Contracts and Liability. Washington, D.C.: National Association of Home Builders, 1987.*

Buying A New Home: A Step-By-Step Guide. Washington, D.C.: National Association of Home Builders, 1985.*

Buying Lots from Developers [HUD-357-I(6)], Washington, D.C.: U.S. Department of Housing and Urban Development, 1982.

A Citizen's Guide to Radon. Washington, D.C.: U.S. Environmental Protection Agency and U.S. Centers for Disease Control, 1986.

Consumer Handbook on Adjustable Rate Mortgages [420-T]. Pueblo, CO: Consumer Information Center, 1984.

Estimating for Home Builders, by John C. Mouton. Washington, D.C.: National Association of Home Builders, 1988.*

Fannie Mae's Consumer Guide to Adjustable Rate Mortgages. Washington, D.C.: Federal National Mortgage Association, 1988.

Financing Land Acquisition and Development. Washington, D.C.: National Association of Home Builders, 1987.*

Home Buying, by Henry S. Harrison. Chicago: National Association of Realtors®, 1980.

---

* Books published by the National Association of Home Builders are available from NAHB Bookstore. For more information, call NAHB Bookstore at 800/368-5242, ext. 463 or 202/822-0463 or write NAHB Bookstore, 15th and M Streets, N.W., Washington, D.C. 20005.

The Home Remodeling Management Book, by Kathryn Schmidt, AIA. Palo Alto, CA: Egger Publications, 1987.

"Homeowner and Architect: Building a Relationship," by Robert L. Miller. *Home*, May 1985.

House, by Tracy Kidder. Boston: Houghton Mifflin Company, 1985.

How to Choose a Remodeler Who's On the Level. Washington, D.C.: National Association of Home Builders, 1987.*

How to Design and Build Your Own Home, by Lupe DiDonno and Phyllis Sperling. New York: Alfred A. Knopf, Inc., 1984.

How to Shop for a Mortgage. Washington, D.C.: Mortgage Bankers Association of America, 1988.

Kitchens, by Jane Moss Snow. Washington, D.C.: National Association of Home Builders, 1987.*

Land Buying Checklist (3rd Edition), by Ralph M. Lewis. Washington, D.C.: National Association of Home Builders, 1988.*

Land Development (7th Edition). Washington, D.C.: National Association of Home Builders, 1987.*

Log Homes Construction and Finance Guidebook. Herndon, VA: Home Buyer Publications, Inc., 1987.

The Mortgage Money Guide [135-T]. Pueblo, CO: Consumer Information Center, 1986.

New Home Buyers Guide. Baltimore: Home Builders Association of Maryland, 1987.

Radon Reduction: A Homeowner's Guide (2nd Edition). Washington, D.C.: U.S. Environmental Protection Agency, 1987.

Radon Reduction in New Construction: An Interim Guide. Washington, D.C.: U.S. Environmental Protection Agency, 1987.

Radon Reduction Techniques for Detached Houses: Technical Guidance. Washington, D.C.: U.S. Environmental Protection Agency, 1986.

Residential Wastewater Systems. Washington, D.C.: National Association of Home Builders, 1980.*

Scheduling for Builders, by Jerry Householder. Washington, D.C.: National Association of Home Builders, 1987.*

Seniors Housing: A Development and Management Handbook. Washington, D.C.: National Association of Home Builders, 1987.*

Settlement Costs [HUD-398-H(3)]. Washington, D.C.: U.S. Department of Housing and Urban Development, 1987.

---

* Books published by the National Association of Home Builders are available from NAHB Bookstore. For more information, call NAHB Bookstore at 800/368-5242, ext. 463 or 202/822-0463 or write NAHB Bookstore, 15th and M Streets, N.W., Washington, D.C. 20005.

Smart House: The Coming Revolution in Housing, by Ralph Lee Smith. Columbia, MD: G.P. Courseware, 1988.*

Understanding Building Codes and Standards in the United States. Washington, D.C.: National Association of Home Builders, 1986.*

Wood Frame House Construction. Armonk, NY: Armonk Press and National Association of Home Builders, 1988.*

You and Your Architect. Washington, D.C.: American Institute of Architects, 1987.

Your New Home and How To Take Care of It. Washington, D.C.: National Association of Home Builders, 1988.*

Organizations

American Arbitration Association
1730 Rhode Island Avenue, N.W.
Washington, D.C. 20036
202/296-8510

American Horticultural Society
Box 0105
Mt. Vernon, VA 22121
703/768-5700

American Institute of Architects
1735 New York Avenue, N.W.
Washington, D.C. 20006
202/626-7300

American Land Title Association
1828 L Street, N.W.
Washington, D.C. 20036
202/296-3671

American Society of Interior Designers
1430 Broadway
New York, NY 10018
212/944-9220

American Society of Landscape
  Architects
1733 Connecticut Avenue, N.W.
Washington, D.C. 20009
202/466-7730

Consumer Information Center
Pueblo, CO 81009

Federal Insurance Administration
500 C Street, S.W.
Washington, D.C. 20472
202/646-2780

Mortgage Bankers Association of America
1125 15th Street, N.W.
Washington, D.C. 20005
202/861-6500

National Academy of Conciliators
5530 Wisconsin Avenue, N.W., Suite 1250
Chevy Chase, MD 20815
301/654-6515

National Association of Home Builders
15th and M Streets, N.W.
Washington, D.C. 20005
202/822-0200

National Kitchen and Bath Association
124 Main Street
Hackettstown, NJ 07840
201/852-0033

---

* Books published by the National Association of Home Builders are available from NAHB Bookstore. For more information, call NAHB Bookstore at 800/368-5242, ext. 463 or 202/822-0463 or write NAHB Bookstore, 15th and M Streets, N.W., Washington, D.C. 20005.